Lecture Notes in Physics

Bisher erschienen/Already published

Vol. 1: J. C. Erdmann. Wärmeleitung in Kristallen, theoretische Grundlagen und fortgeschrittene experimentelle Methoden. II, 283 Seiten. 1969.

Vol. 2: K. Hepp, Théorie de la renormalisation. III, 215 pages. 1969.

Vol. 3: A. Martin, Scattering Theory: Unitarity, Analyticity and Crossing. IV, 125 pages. 1969.

Vol. 4: G. Ludwig, Deutung des Begriffs „physikalische Theorie" und axiomatische Grundlegung der Hilbertraumstruktur der Quantenmechanik durch Hauptsätze des Messens. 1970. Vergriffen.

Vol. 5: Schaaf, The Reduction of the Product of Two Irreducible Unitary Representations of the Proper Orthochronous Quantummechanical Poincare Group. IV, 120 pages. 1970.

Vol. 6: Group Representations in Mathematics and Physics. Edited by V. Bargmann. V, 340 pages. 1970.

Vol. 7: R. Balescu, J. L. Lebowitz, I. Prigogine, P. Résibois, Z. W. Salsburg, Lectures in Statistical Physics. V, 181 pages. 1971.

Vol. 8: Proceedings of the Second International Conference on Numerical Methods in Fluid Dynamics. Edited by M. Holt. 1971. Out of print.

Vol. 9: D. W. Robinson, The Thermodynamic Pressure in Quantum Statistical Mechanics. V, 115 pages. 1971.

Vol. 10: J. M. Stewart, Non-Equilibrium-Relativistic Kinetic Theory. III, 113 pages. 1971.

Vol. 11: O. Steinmann, Pertubation Expansions in Axiomatic Field Theory. III. 126 pages. 1976.

Vol. 12: Statistical Models and Turbulence. Edited by C. Van Atta and M. Rosenblatt. Reprint of the First Edition. VIII, 492 pages. 1975.

Vol. 13: M. Ryan, Hamiltonian Cosmology. VII, 169 pages. 1972.

Vol. 14: Methods of Local and Global Differential Geometry in General Relativity. Edited by D. Farnsworth, J. Fink, J. Porter, and A. Thompson. V, 188 pages.

Vol. 15: M. Fierz, Vorlesungen zur Entwicklungsgeschichte der Mechanik. V, 97 Seiten. 1972.

Vol. 16: H.-O. Goergii, Phasenübergang 1. Art bei Gittergasmodellen. IX, 167 Seiten. 1972.

Vol. 17: Strong Interaction Physics. Edited by W. Rühl and A. Vancura. V, 405 pages. 1973.

Vol. 18: Proceedings of the Third International Conference on Numerical Methods in Fluid Mechanics, Vol. I. Edited by H. Cabannes and R. Temam. VII, 186 pages. 1973.

Vol. 19: Proceedings of the Third International Conference on Nemerical Methods in Fluid Mechanics, Vol. II. Edited by H. Cabannes and R. Temam. VII, 275 pages. 1973.

Vol. 20: Statistical Mechanics and Mathematical Problems. Edited by A. Lenard. VIII, 247 pages. 1973.

Vol. 21: Optimization and Stability Problems in Continuum Mechanics. Edited by P. K. C. Wang. V, 94 pages. 1973.

Vol. 22: Proceedings of the Europhysics Study Conference on Intermediate Processes in Nuclear Reactions. Edited by N. Cindro, P. Kulišic and Th. Mayer-Kuckuk. XIV, 329 pages. 1973.

Vol. 23: Nuclear Structure Physics. Proceedings 1973. Edited by U. Smilansky, I. Talmi, and H. A. Weidenmüller. XII, 296 pages. 1973.

Vol. 24: R. F. Snipes, Statistical Mechanical Theory of the Electrolytic Transport of Non-electrolytes. V, 210 pages. 1973.

Vol. 25: Constructive Quantum Field Theory. The 1973 "Ettore Majorana" International School of Mathematical Physics. Edited by G. Velo and A. Wightman. III. 331 pages. 1973.

Vol. 26: A. Hubert, Theorie der Domänenwände in geordneten Medien. XII, 377 Seiten. 1974.

Vol. 27: R. K. Zeytounian, Notes sur les Ecoulements Rotationnels de Fluides Parfaits. XIII, 407 pages. 1974.

Vol. 28: Lectures in Statistical Physics. Edited by W. C. Schieve and J. S. Turner. V. 342 pages. 1974.

Vol. 29: Foundations of Quantum Mechanics and Ordered Linear Spaces. Advanced Study Institute, Marburg 1973. Edited by A. Hartkämper and H. Neumann. VI, 355 pages. 1974.

Vol. 30: Polarization Nuclear Physics. Proceedings 1973. Edited by D. Fick. IX, 292 pages. 1974.

Vol. 31: Transport Phenomena. Sitges International Schools of Statistical Mechanics, June 1974. Edited by G. Kirczenow and J. Marro. XIV, 517 pages. 1974.

Lecture Notes in Physics

Edited by J. Ehlers, München, K. Hepp, Zürich,
H. A. Weidenmüller, Heidelberg, and J. Zittartz, Köln
Managing Editor: W. Beiglböck, Heidelberg

52

Milorad Mladjenović

Development of Magnetic β-Ray Spectroscopy

Springer-Verlag
Berlin Heidelberg GmbH 1976

Author
Milorad Mladjenović
Boris Kidric Institute of Nuclear Sciences
P.O.Box 522
11001 Beograd/Jugoslavia

Library of Congress Cataloging in Publication Data

Mladenović, Milorad, 1920-
    Development of magnetic B-ray spectrometers.

    (Lecture notes in physics ; 52)
    Includes bibliographical references.
    1. Spectrometer. 2. Beta ray spectrometry.
I. Title. II. Series.
QC373.S7M55   539.7'2112   76-26517

ISBN 978-3-540-07851-7   ISBN 978-3-540-38199-0 (eBook)
DOI 10.1007/978-3-540-38199-0

Originally published by Springer-Verlag Berlin Heidelberg New York in 1976

To
Nobel Institute for Physics
Birthplace of Modern Beta-Ray Spectroscopy

FOREWORD

Although beta-ray spectroscopy is about 65 years old, there is no monograph, or any other kind of a book entirely devoted to a complete exposition of design and construction of beta-ray spectrometers. Development has been covered periodically only by review articles in books and journals. The situation is completely opposite from the one in the neighbouring field of electron microscopy, which abounds in all kinds of monographs and handbooks. That is the main reason why I have tried to include as many as possible different types of spectrometers which were developed in the past, although quite a few of them might not be used now for the measurement of beta spectra. No spectroscopy has ever died. It may be neglected in one field to start a new life in another. They are often born in physics, pass to chemistry and migrate to many applied fields, from metallurgy to archeology. Since some of those new uses might not need the sophisticated devices developed in nuclear physics, a rather complete coverage of various types can help to find the optimum design for a given case.

This book was developed gradually through post-graduate courses given in Belgrade, Rome, Cairo and Nashville. I am grateful to Prof. S. Sciuti, M. El-Nadi and J. H. Hamilton for the hospitality and the chance given to write up some lectures which are parts of this book.

My initiation in this field started in 1948, when Dr. R. J. Walen, the head of the Physics laboratory gave me the subject for my diploma work: "Le calcul des trajectoires des électrons dans une lentille magnétique." His tireless guidance prepared me for the next phase, which was the work in the Nobel Institute for Physics, Stockholm. It was a bit of luck to come to Stockholm just at the time when they had completed the first large double-focusing spectrometer and the whole field was in the initial stages of the well-known boom. I cannot forget the spirit and people I met there, and I remain forever grateful to Profs. Manne Siegbahn, Kai Siegbahn, Hilding Slätis, Arne Hedgran and Ingmar Bergström.

Once back in Belgrade, I was given the chance by my professor, Pavle Savić, founder of "B. Kidrič" Institute, to build several spectrometers. I am grateful to him for bringing me to this Institute and giving me all the opportunities that a young man could wish.

My thanks are due to Mr. Nikola Skorupan, who not only helped me constructing the spectrometers, but also made the drawings for this book.

Belgrade, December 1974   Milorad Mladjenović

CONTENTS

1. HISTORICAL INTRODUCTION

The use of magnetic fields to study the properties of charged particles dates back to the early investigation of these particles. The deflection of a particle in a magnetic field served sometimes as first indication, and always as a safe proof, that it is charged.

First, the corpuscular radiations produced in gas discharges were studied. Later, with the discovery of radioactivity, the application of magnetic fields showed that radiations emitted by radioactive substances consist of three different kinds, two of them, called alpha- and beta- rays, being deflected by a magnetic field in opposite directions, while the third, called gamma-rays, left unaffected by a field.

1910 - 1950

The first systematic study of beta-rays by magnetic analysis was performed by von Baeyer, Hahn and Meitner (1,2). They constructed a spectrometer in which a very narrow beam of beta-rays was deflected by a homogenous magnetic field, and then detected by a photogaphic plate. (Fig.1.1). They found that the spectra contained several lines superimposed on a continuous background. From the positions of the source, slit and the lines, and knowing the magnetic field, it was possible to find the momenta of various electron groups.

A decisive advance was made when Danysz showed (3) that when the angle of deflection is increased to 180°, a focusing of the beam is obtained, which greatly improves the intensity of the lines and separation between them. This method was then taken up and developed by Rutherford and Robinson (4) into a type of spectrometer, called semi-circular. (Fig.1.2) Several variations of design of this type of spectrometer, were gradually developed. In the beginning, the magnetic field was produced by electromagnet, and later permanent magnets were introduced by Ellis, Cockroft and Kershaw (5), collaborators of Rutherford. Chadwick, also from Cavendish, was the first to use a particle counter (6). The semi-circular spectrometers were the principal type in use up to the World War II.

Beginnig with 1924, another, quite different type of spectrometer was developed. The magnetic field was produced by a coil, as shown in Fig.1.3. A conical beam leaves the source placed at the field symmetry axis, returns and converges to another position on the axis, in a manner similar to the focusing of light rays by an optical lens. The magnetic lens spectrometer was suggested by Kapitza (7) and built first by Tricker (8), who used a long coil producing a uniform field, and later Klemperer constructed the first short magnetic lens (9). In a magnetic lens, compared to semi-circular type, a larger part of radiation is transmitted to the detector, but the separation between the lines is generally poorer. As the spectra of naturally radioactive nuclei are usually quite complex and contain many lines, high resolving power spectrometers are needed for their measurements. For this reason, magnetic lenses, although discovered in 1924, did not come into widespread use before a more intensive study began of artificially produced radioactive lsotopes, which initially were often produced with weaker activities, and sometimes had much simpler spectra. The magnetic lens was, however, extensively used and studied in the electron microscopy.

Around 1940 several laboratories began buiding magnetic lens spectrometers. Witcher constructed one with a uniform field in Columbia University (10), and Deutch, Elliot and Evans from M.I.T. published a detailed description of a short lens (11). At Nobel Institute, in Stockholm, Kai Siegbahn started developing lenses which culminated in the construction, by Slatis and himself, of the "intermediate image", a lens with very high transmission and small image size (12-14). Subsequently, most of the laboratories working beta spectroscopy built magnetic lenses of various design.

The semi-circular and lens spectrometers represent the simplest types with modest performances. For further improvements, special magnetic fields were needed. Thus, semi-circular type was improved in two different manners. Beiduck and Konpinski (15) caculated the field which corrects third--order aberrations due to radial aperture and two such spectrometers were built (16-17). On the other hand, several groúps from Leningrad (18-22) improved the focusing by combining homogenous with one-dimensionally decreasing field (ketron).

1950 – 1960

The most important advance took place, however, with the invention of two new types, the double-focusing and the toroidal spectrometers. The first proposal for the double-focusing spectrometer was put forward by Siegbahn and Swartholm (23-24). They made first a small model and by 1950 a large spectrometer was built in Nobel Institute (25). Fig.1.4a. The difference between this instrument and all the others made before was so important that it marks the beginning of a new, second phase in the development of beta-spectrometers. It had the advantage over the semi-circular of two-dimentional focusing, twice larger dispersion, while using a larger source. Another important break-through of Nobel Institute instrument was it's size. While most of the beta-spectrometers made previously had diameters of pole-pieces between 20-40cm, it had diameter of 160 cm and the radius of mean electron orbit was 50 cm.

Many double-focusing spectrometers were later built in other laboratories. In the beginning, the magnetic field was produced by shaping the iron pole-pieces and later Siegbahn and Edwardson (27) (Fig.1.4b) and Moussa and Bellicard (28) used special combinations of coils for construction of iron-free spectrometers.

The toroidal spectrometer was developed independently in several places. After Harris proposed the construction of a toroidal spectrometer for cosmic rays in 1947 (27), Szalay suggested it could also be used for beta spectroscopy and Horvath made some initial calculations (30). The first detailed theoretical treatment was published by Kofoed-Hansen, D. B. Nielsen and Linhard (31). Subsequently, several toroidal spectrometers were built in Denmark (32-33). (Fig.1.5a). The toroidal magnetic field is produced by a series of iron sectors placed symmetrically around an axis, which contains the source and the detector. There is also an iron-free version first developed in Moscow by Vladimirskii, Tarasov and Trebuhovskii (34), in which the field is produced by a toroidal coil. It was later further developed by several groups in the same laboratory (35-36) (Fig.1.5b).

The main advantage of the toroidal spectrometer is very high transmission, which so far has not been attained by any other type. The resolution is comparatively good and it is also convenient to have the source and the detector outside of the field.

In the decade 1950 - 1960 several new types of beta spectrometers were developed, besides the double-focusing and toroidal. The largest contribution is due to Leningrad beta spectroscopists. Dzhelepov and his collaborators developed a new type of electron spectrometer specialized for the measurements of photons. It was first called ritron, and later, an improved version, elotron (37–40). The basic idea is to let photons hit a large radiator and collect Compton electrons into an intermediate focus, which serves as the source for another magnetic analyser (Fig. 1.6). This offers the possibility to improve the efficiency and the resolution, and also reducing the background, which is always a problem when Compton lines serve to measure the photon spectra. The same type was further developed and adapted for the measurements of neutron capture gamma--rays, bv Groshev and collaborators in Moscow (41). The idea of using large radiator and intermediate focus proved to be a successfull one, especially since many different combinations of fields are possible. Later, Groshev and Demidov combined two special fields and obtained a substantial improvement of resolution and efficiency (42). After the development of Ge(Li) counters, the use of Compton magnetic spectrometers declined.

Another new type was developed in Leningrad by Kelman and collaborators (43–45). The spectrometer consists of two lenses and a deflecting magnet between them, in analogy to optical spectrographs (Fig. 1.7). It is a large apparatus with which it is possible to achieve very high dispersions.

The late 1950's mark the end of what may be called artisan approach to spectrometer construction. They cease to be a one-man affair and become an engineering project, requiring team work and an order of magnitude higher expences. The examples are one meter radius iron-free double focusing spectrometer in Chalk River (46) and large iron-free toroidal spectrometer in Argonne (47).

1960 - 1970

Many cases of unsolved complex spectra required still higher transmissions and resolutions. Several approaches were tried to improve the performances.

One possible approach was to improve existing spectrometers by additional electric or magnetic fields. Bergkvist used electric fields to correct for aberrations and to increase the source area (48). In this way he succeeded in increasing the luminosity up to two orders of magnitude. A series of different correctors was developed.

Another approach was to use the possibilities offered by cylindrical fields that by increasing the angle of radial focusing, an important improvement of dispersion may be achieved. The disadvantage is that the focusing angle may become greater than 360°, so that the electrons pass the detector and the source before being focused. Spectrometers of this type were developed by Daniel and Jahn (49) and by Baranov and his collaborators (50) (Fig.1.8).

Although the idea of trochoidal electron paths was introduced by Thibeaud in early thirties, the first spectrometers with good performances were being calculated and made in late sixties. Two groups, one in Lion, led by Lafoucriere and another in Zurich, led by Hofman, made extensive calculations and each built a spectrometer (51-53). (Fig.1.9).

A novel approach consists in giving up the rotational symmetry, in order to obtain new degrees of freedom to deal with aberrations. Only initial theoretical work was done so far, by Sessler and Bergkvist(51-55) and by Daniel and Schmutzler (56). They have demonstrated that fields can be found with good focusing characteristics. The corresponding magnets are not simple to make and no designs were proposed yet.

References Ch.1

1.   O.von Baeyer and O.Hahn, Phys.Zeit. 11, 488 (1910)

2.   O.von Baeyer, O.Hahn and L.Meitner, Phys.Zeit. 12, 273 and 378 (1911)

3.   J.Danysz, C.R.Acad.Sci. Paris, 153, 339 and 1036 (1911) Le Radium, 9, 1 (1912) and 10, 4 (1913)

4.   E.Rutherford and H.Robinson, Phil.Mag. (6) 26, 717 (1913)

5.   J.D.Cockroft, C.D.Ellis and H.Kershaw, Proc.Roy.Soc.A132, 442 (1931)

6.   J.Chadwick, Verh. Deutch. Phys. Ges. 16, 383 (1914)

7.   P.Kapitza, Proc. Cambr. Phil. Soc. 22, 3 (1925)

8.   R.Tricker, Proc. Cambr. Phil. Soc. 22, 454 (1925)

9.   O.Klemperer, Phil. Mag. 20, 545 (1935)

10. C.M.Witcher, Phys. Rev. 60, 32 (1941)

11. M.Deutsch, L.G.Elliot and R.D.Evans, Rev. Sci. Instr. 15, 118 (1944)

12. K.Siegbahn, Ark. Mat. Astron. Fys. Ser. A, 28, No.17 (1942)

13. K.Siegbahn, Ark. Mat. Astron. Fys. Ser. A, 30, No.1 (1943)

14. H.Slatis and K.Siegbahn, Ark. Fysik 1, 339 (1949)

15. F.M.Beiduk and E.J.Konopinski, Rev. Sci. Instr. 19, 594 (1948)

16. L.M.Langer and C.S.Cook, Rev. Sci. Instr. 19, 257 (1948)

17. J.A.Bruner and F.R.Scott, Rev. Sci. Instr. 21, 545 (1950)

18. M.Korsunskii, V.Kelman and B.Petrov, Zhurn. Eksp. Teor. Fiz. 14, 394 (1944)

19. B.S.Dzhelepov and A.A.Bashilov, Izvestia AN, Ser. Fiz. 14, 263 (1950)

20. P.P.Pavinskii, Izvestia AN, Ser. Fiz. 18, 175 (1954)

21. A.A.Bashilov and V.I.Bernotas, Izvestia AN, Ser. Fiz. 18, 192 (1954)

22. G.D.Latishev, A.G.Sergeyev, Y.M.Krisyuk, L.A.Ostrecov, Y.S. Yegorov and N.M.Shirshov, Izvestia AN, Ser. Fiz. 20, 354 (1956)

23. N.Svartholm and K.Siegbahn, Ark. Mat. Astron. Fys. Ser. A, 33, No.21 (1946)

24. K.Siegbahn and N.Svartholm, Nature, Lond. 157, 872 (1946)

25. A.Hedgran, K.Siegbahn and N.Svartholm, Proc. Phys. Soc. Lond. Ser. A 63, 960 (1950)

26. A.Hedgran, Ark. Fysik 5, 1 (1951)

27. K.Siegbahn and K.Edvardson, Nucl. Phys. 1, 137 (1956)

28.   A. Moussa and J. B. Bellicard, J. Phys. Rad. **15**, 85A (1954)

29.   W. T. Harris, Phys. Rev. **71**, 310 (1947)

30.   J. Horvath, Experientia **5**, 112 (1949)

31.   O. Kofoed - Hansen, J. Linhard and O. B. Nielsen, Mat. Fys. Medd. Dan.
       Vid. Selsk. **25**, No.16 (1950)

32.   O. B. Nielsen and O. Kofoed - Hansen, Mat. Fys. Medd. Dan. Vid. Selsk.
       29, No.6 (1955)

33.   K. M. Bisgard, Nucl. Instr. Meth. **22**, 221 (1963)

34.   V. V. Vladimirskii, E. K. Tarasov and Yu. V. Trebuhovskii, Prib. Tek.
       Eksp. No.1, 13 (1956)

35.   Eu. F. Tretyakov, L. L. Goldin and G. I. Grishuk, Prib. Tek. Eksp. No.6,
       22 (1957)

36.   N. A. Burgov, A. V. Davidov and G. R. Kartashov, Nucl. Instr. Meth. **12**,
       316 (1961)

37.   B. S. Dzhelepov and M. L. Orbeli, Dokladi AN SSSR, Ser. Fiz. **62**, 615
       (1948)

38.   B. S. Dzhelepov, Dokladi AN SSSR, Ser. Fiz. **77**, 383 (1951)

39.   B. S. Dzhelepov, N. Zhukovskii and Y. Holnov, Izvestia AN SSSR, Ser. Fiz.
       18, 599 (1954)

40.   B. S. Dzhelepov and S. A. Shestopalova, Izvestia AN SSSR, Ser. Fiz. **20**,
       328 (1956)

41.   L. V. Groshev, B. P. Adiyashevich and A. M. Demidov, Proceed. Geneva
       Conf. 1955, vol.2, p.39

42.   L. V. Groshev, A. M. Demidov, V. N. Lutsenko and A. F. Malov, Izvestia
       AN SSSR, Ser. Fiz. **24**, 791 (1960)

43.   V. M. Kelman and D. L. Kaminskii, Zhur. Eksp. Teor. Fiz. **21**, 555 (1951)

44.   V. M. Kelman, D. L. Kaminskii and V. A. Romanov, Izvestia AN SSSR, Ser.
       Fiz. **18**, 148 (1954) and **18**, 209 (1954)

45.   V. M. Kelman, B. P. Peregud and V. I. Skopina, Nucl. Instr. Meth. **27**, 190
       (1964)

46.   R. L. Graham, G. T. Ewan and J. S. Geiger, Nucl. Instr. Meth. **9**, 245
       (1960)

47.   M. S. Freedman, F. Wagner Jr., F. T. Porter, J. Terandy and P. P. Day,
       Nucl. Instr. Meth. **8**, 255 (1960)

48.   K. E. Bergkvist, Ark. Fys. **27**, 383 and 439 (1964); Nucl. Instr. Meth.
       43, 170 (1966)

49.   H. Daniel, P. Jahn, M. Kunze and G. Spanagel, Nucl. Instr. Meth. **35**, 171
       (1965)

50.   S.A.Baranov, R.M.Polevoi, Yu.A.Aliev and S.N.Belenkii, Prib. Tek. Eksp. No.6, 64 (1965)

51.   J.Lafoucriere, Ann. Phys. 6, 610 (1951)

52.   G.Lucenet, Nucl. Instr. Meth. 24, 51 (1963)

53.   R.Balzer, D.Barucha, F.Heinrich and H.Hofman, Nucl.Instr. Meth. 57, 277 (1967)

54.   A.M.Sessler, Nucl. Instr. Meth. 23, 165 (1963)

55.   K.E.Bergkvist and A.M.Sessler, Nucl. Instr. Meth. 46, 317 (1967)

56.   F.Schmutzler and H.Daniel, Nucl. Instr. Meth. 83, 13 (1965)

Text to Figures. Ch.1

Fig. 1.1.   The first spectrometer. S – source, P – photographic plate, B – slit. (1.2)

Fig. 1.2.   The first semi–circular spectrometer. (4)

Fig. 1.3.   The first lens spectrometers. (a) long lens (8), (b) short lens (9)

Fig. 1.4.   (a) The first large iron-core double – focusing spectrometer (25)

Fig. 1.4.   (b) The first iron – free double – focusing spectrometer (27)

Fig. 1.5.   (a) The first iron – core toroidal beta spectrometer (32)

Fig. 1.5.   (b) The iron – free toroidal spectrometer (36)

Fig. 1.6.   The elotron (40)

Fig. 1.7.   The optical analogy spectrometer (43)

Fig. 1.8. $(\pi/2)\sqrt{13}$ spectrometer. (a) Beam geometry; (b) coil geometry (49)

Fig. 1.9.   The 6–loop trochoidal spectrometer (53)

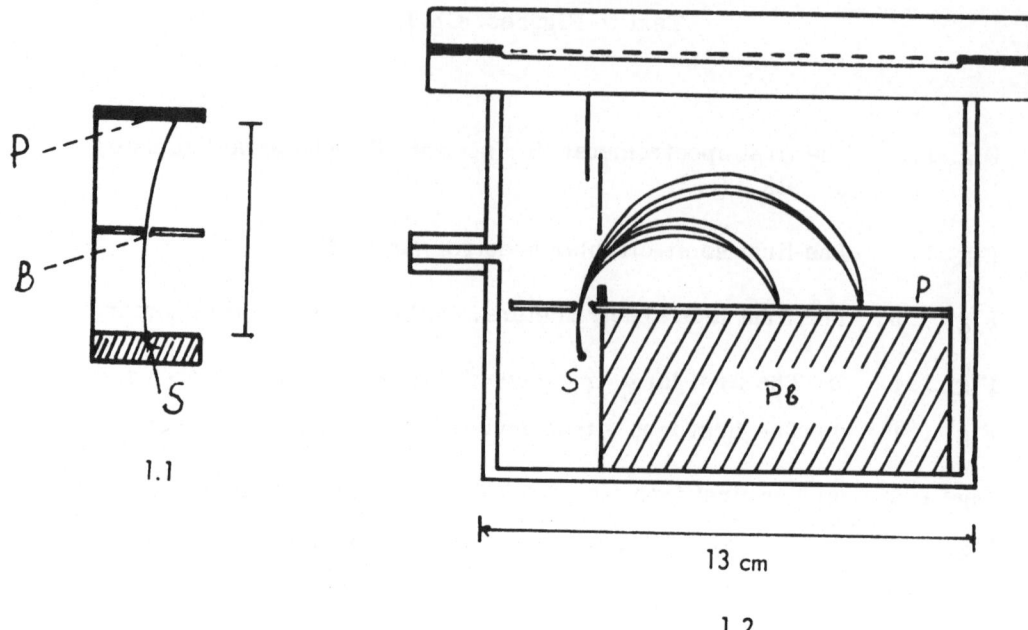

1.1

1.2

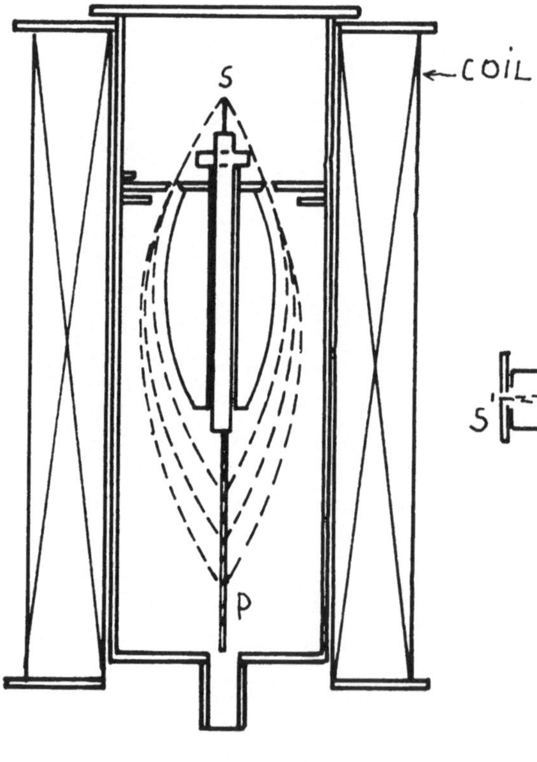

13 cm

(a)

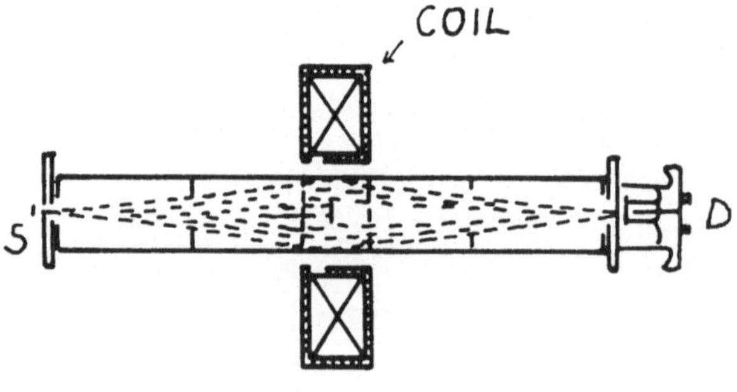

(b)

1.3

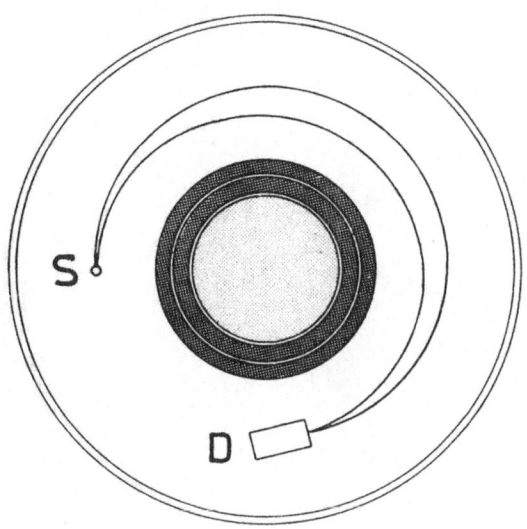

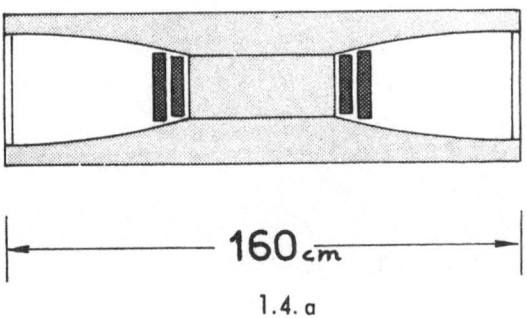

S

D

160 cm

1.4. a

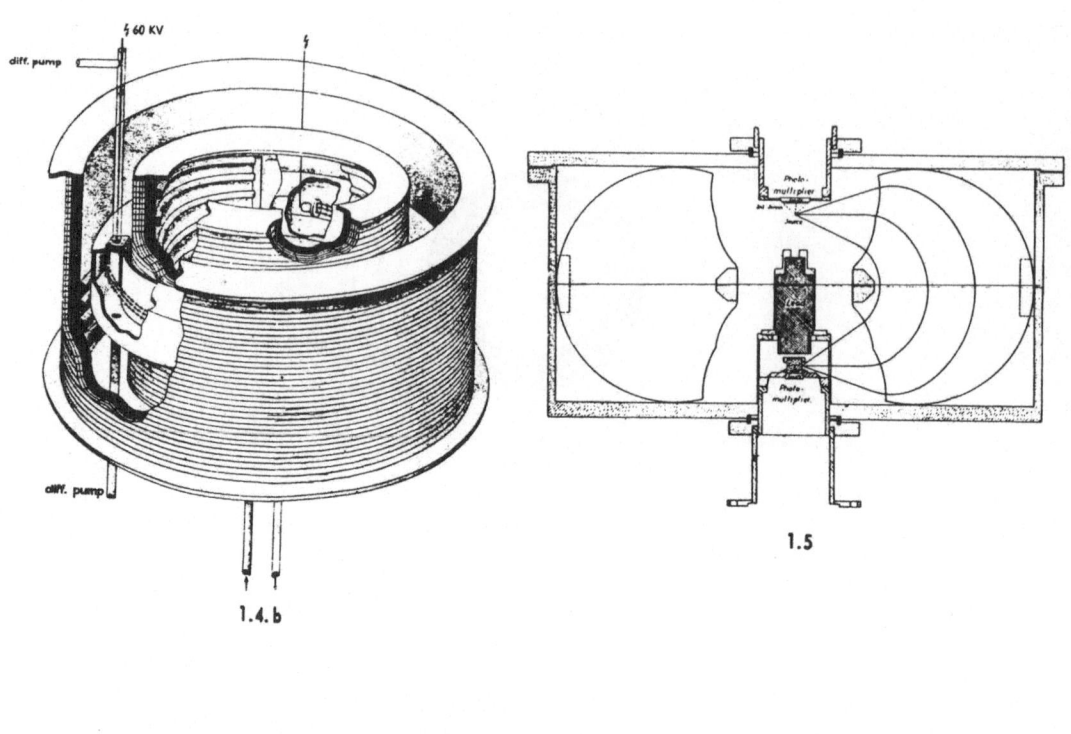

1.4. b

1.5

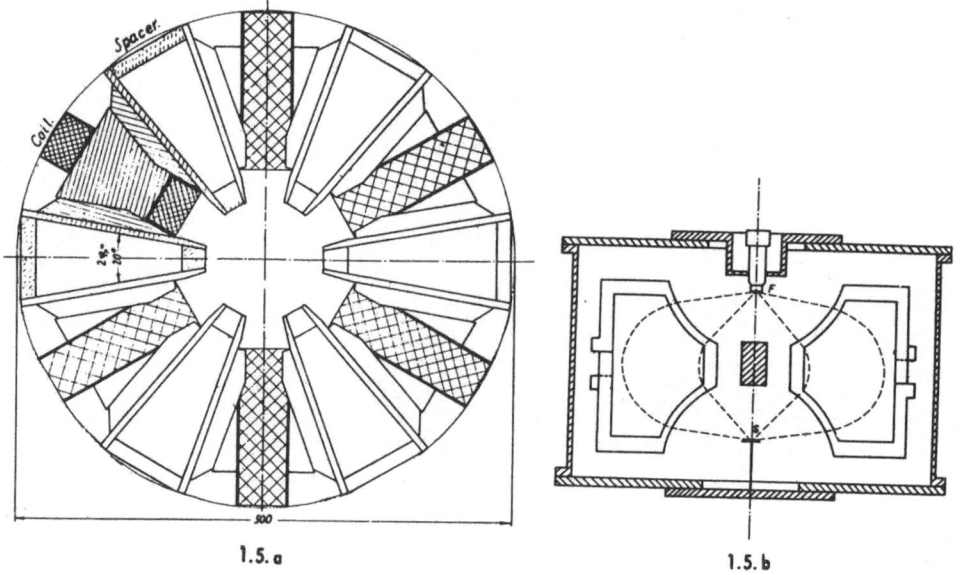

1.5. a

1.5. b

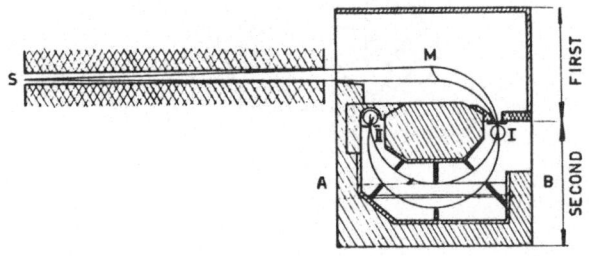

1.6

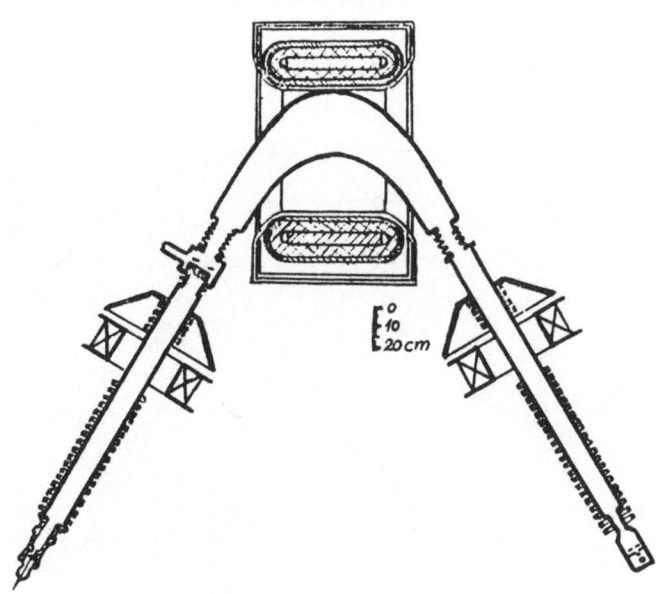

1.7

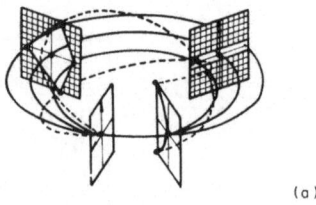

(a)

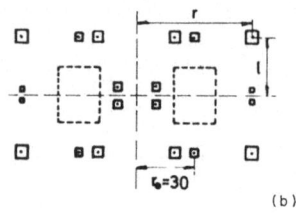

(b)

1.8

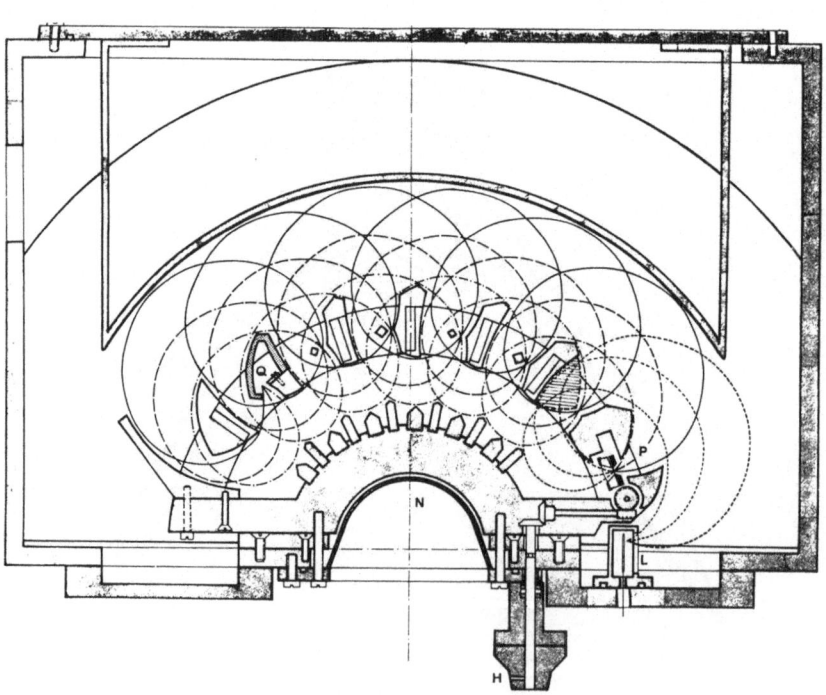

1.9

2. ELEMENTARY DESCRIPTION OF SPECTROMETERS

The spectrum of beta particles can be measured by magnetic, electric and electro-magnetic fields. Since the energies of beta particles may amount to a few MeV, unconveniently high electric potentials are needed if electric fields are used. Another disadvantage of the electrostatic field is that its focusing equation is not relativistically invariant, so that the parameters of the focus change when going from small to relativistic energies. For these reasons the electrostatic beta-ray spectrometers are not much used. Recently electrostatic electron spectrometers are being built for measurements of low energy electrons in ESCA work.

Due to relativistic velocities of beta-particles, time varying fields are of no value. Magnetostatic fields are, therefore, the most convenient for momentum analysis of beta particles and in the following, we shall deal only with magnetostatic spectrometers.

2.1. Equations of motion

The force f acting on an electron, moving in an electromagnetic field is given by Lorentz equation

$$\vec{f} = e\,(\vec{E} + \frac{1}{c}\,\vec{v} \times \vec{B}) \tag{2.1}$$

where $\vec{E}$ is the electric field, $\vec{B}$ is the magnetic induction, and $\vec{f}$ is called Lorentz force.

It there is no electric field, Lorentz force reduces to

$$\vec{f}_{dyne} = e_{emu}\,\vec{v}_{cmsec^{-1}} \times \vec{B}_{gauss} \tag{2.2}$$

expressed in c.g.s electromagnetic units.

The same equation in M.K.S. system is written as

$$\vec{f}_{newtons} = e_{coulomb}\,\vec{v}_{metres} \times \vec{B}_{weber\ per\ m^2}$$

with the following conversion ratios

$$1 \text{ newton} = 10^5 \text{ dynes}$$

$$1 \text{ weber per } m^2 = 10^4 \text{ gauss}$$

and $\qquad e = 1.6 \times 10^{19} \text{ coulomb}$

In the following, the c.g.s. electromagnetic units shall be used.

The Lorentz force acting on an electron is always perpendicular to the velocity vector, so that only direction of velocity is changing. The magnitude of velocity remains constant, which means that energy also remains constant in a magnetostatic field.

The equation of motion is

$$\frac{d\vec{p}}{dt} = e\,\vec{v} \times \vec{B} \tag{2.3}$$

or writing relativistic momentum p explicitly

$$\frac{d}{dt}\,\frac{m_o \vec{v}}{(1-v^2/c^2)^{1/2}} = e\,\vec{v} \times \vec{B} \tag{2.4}$$

2.2.. The Radius of Curvature of Electron Trajectory in a Magnetic Field

The Lorentz force, being always perpendicular to the velocity vector, produces a curvature, which can be quantitatively determined by taking as reference system the tangent $\vec{s}_1$ normal $\vec{n}$ and binormal $\vec{b}$ as unit vectors at a given point of trajectory (Fig.2.1). The accelaration is then given by the well-known relation

$$\frac{d\vec{v}}{dt} = \vec{s}\,\frac{dv}{dt} + \frac{v^2}{\rho}\,\vec{n} \tag{2.5}$$

where ρ is the radius of curvature. Differentiating left side of (2.4) and using (2.5), the equation of motion becomes

$$m_o(1 - v^2/c^2)^{-1/2}\,v^2/\rho\,\vec{n} + m_o(1 - v^2/c^2)^{-3/2}\,\frac{dv}{dt}\,\vec{s} = e\vec{v} \times \vec{B} \tag{2.6}$$

Multiplying by normal unit vector $\vec{n}_1$ (2.6) reduces to

$$m_o(1 - v^2/c^2)^{-1/2}\,v^2/\rho = e\,B_b\,v \tag{2.7}$$

where B_b is the component of the magnetic field along the binormal.

Rearranging, (2.7) becomes

$$(m_o v/e)(1 - v^2/c^2)^{-1/2} = B_b \rho \tag{2.8}$$

or $$p/e = B_b \rho \tag{2.9}$$

In the simple case of electron moving in a plane perpendicular to a homogenous magnetic field, $B_b = B$ is constant and ρ must also be constant. The electron is then desribing a circle having ρ as radius. The electron momentum is proportional to the product of magnetic field and the radius of the circle, representing the electron trajectory, in that field (supposed constant over the circle and perpendicular to the plane containing it). Precisely these two quantities, the magnetic field and orbit radius, are in most cases measured in a magnetic spectrometer (or if one of them remains constant during the experiment, its value is known), so that the results of measurements give directly the momenta of electrons. It is convenient then to express the momenta in $B\rho$ units. Since the decay schemes are discussed in terms of particle energies, the tables are needed for conversion between energy expressed in Ev and momentum expressed in $B\rho$ gauss cm. The relation between them is derived below.

In the M.K.S. system $B\rho$ is measured in $\dfrac{\text{weber}}{m^2} \times m = \text{weber} \times m^{-1}$, so that the conversion ratio is $B\rho$ gauss.cm $= 10^{-6}$ $B\rho$ weber m^{-1}.

2.3.   Energy-momentum Relations

The relativistic formula for the kinetic energy T of electron is given by.

$$T = m_o c^2 \left| (1 - v^2/c^2)^{-1/2} - 1 \right| \tag{2.10}$$

The expression for the momentum p

$$p = m_o v(1 - v^2/c^2)^{-1/2}$$

can be transformed to

$$(1 - v^2/c^2)^{-1} = (p/c m_o)^2 + 1 \tag{2.11}$$

The relation between kinetic energy and momentum is obtained by combining (2.10) and (2.11)

$$T = m_o c^2 \left\{ \left| (p/cm_o)^2 + 1 \right|^{1/2} - 1 \right\} \tag{2.12}$$

Expressing p in $B\rho$ units with the help of (2.9), energy T becomes

$$T = m_o c^2 \left\{ \left| (e\,B\rho/m_o c)^2 + 1 \right|^{1/2} - 1 \right\} \tag{2.13}$$

The values of constants are:

$$m_o c^2 = 510.976 \text{ keV}$$

$$e/m_o = 1.7589 \times 10^7 \text{ emUg}^{-1}$$

$$c = 2.99793 \times 10^{10} \text{cmsec}^{-1}$$

which introduced into (2.13) give

$$T \text{ (keV)} = 510.976 \left\{ 344.2 \times 10^{-10} (B\rho)^2 + 1 \right.^{1/2} -1 \} \tag{2.13a}$$

This relation is graphically shown in Fig.2.2.

By differentiating (2.13) one obtains

$$\frac{dT}{T} = \frac{T + 2 m_o c^2}{T + m_o c^2} \frac{d(B\rho)}{(B\rho)} = \left| 1 + \frac{510.98}{510.98 + T(\text{keV})} \right| \frac{d(B\rho)}{B\rho} \tag{2.14}$$

which is useful for converting the line widths between momentum and energy units. The relation (2.14) is shown graphically in Fig.2.3. It is good to remember that

$$\frac{d(B\,\rho)}{B\,\rho} < \frac{dT}{T}$$

For instance when one wants to estimate whether two close lines could be separated in the spectrometer, one starts with their relative separation in energy units, and then using the relation (2.14) obtains $d(B\rho)/(B\rho)$. At 400 keV, $d(B\rho)/(B\rho)$ is 40% smaller than dT/T.

2.4. Radioactive Source

Beta-ray spectrometer may serve either for the measurement of electrons emitted directly from a radioactive source or for the momentum analysis of electrons ejected by photons bombarding a radiator. Let us briefly describe the main characteristics of these two types of sources.

A beta-ray spectrum would ordinarily consist of a continuous part, on which some lines may be superimposed. In the measurement of continuous beta spectrum, there are two important magnitudes to be determined: the momentum of the end point p_{max}, and the shape of the spectrum. It is much easier to find p_{max} than the shape of the spectrum, because several factors may contribute to the distortion of the shape. One of these is the source thickness which has to be so small that electrons do not loose any appreciable amount of energy in collisions with the atoms of the source itself, on their way out, since any energy loss produces a shift of the spectrum to the lower momentum side and an accumulation of lower momentum electrons (Fig. 2.4). In order to avoid this type of distortion it is necessary to have uniform source with a thickness of $\leq 10 \, \mu g/cm^2$ A distortion of the continuous spectrum may also be produced by the source-backing, a foil on which the source is spread. Some of the electrons emitted in the direction of the foil experience several collisions and return with a reduced energy back into the beam accepted by spectrometer diaphragmes (Fig. 2.5). This effect can be reduced to negligible proportions by using a source backing of low Z and thickness $< 10 \, \mu g/cm^2$. The scattering of electrons from objects near the source and the walls of vacuum chamber may also increase the number of lower momentum electrons. This is reduced by not having any objects near the source and placing several diaphragmes so that the electrons leaving the wall cannot reach the detector. The walls and other devices inside the vacuum chamber are made from materials with low Z (aluminum and lucite) to reduce the scattering and also to reduce the number of secondary electrons ejected by gamma-rays which accompany the beta-decay. Finally, it should be mentioned that charging of the source, due to the electrons leaving it, may produce the momentum shift which requires that the source holder should be electrically conducting and connected to ground (Fig. 2.6). If precautions are taken to avoid the distortions, the continuous beta spectrum can be measured with almost any type of beta-ray spectrometer and no special performances are required. Often it is very useful to have a gamma spectrometer (usually a Na(TL) or Ge-Li crystal) coupled in coincidence with a beta-spectrometer because beta spectrum is usually complex, consisting of several transitions followed by different gamma rays. Beta-gamma coincidences show which nuclear energy levels are directly populated by beta decay.

The most important part of the line spectrum consists of internal conversion lines. At lower energies (< 80 keV) usually an Auger spectrum also appears. The line spectrum may be very complex containing sometimes more than 100 lines, some lying very close to each other. The conversion lines have a definite width and an asymmetrical shape. Every line has a tail on the low energy side produced mainly by the energy loss in the source and by backscattering from source backing. Besides these asymmetrical contributions to the width of the line there are others due to the dimensions of the source and slits, aberrations and the inherent widths of the atomic levels. All of these shall be considered in more detail later. The resolution of two adjacent lines depend on their widths and relative distance. The ability of a spectrometer to increase the distance between the doublet is called dispersion (more precisely defined in 2.5 and 2.9). Complex internal conversion spectra require spectrometers with high dispersion and resolving power.

A radioactive source emits electrons isotropically into 4π angle. In a beta-spectrometer only a small part of 4π is accepted. Expressed as a percentage of 4π the aperture of the electron beam accepted by the spectrometer is of the order of 1%. Some spectrometers accept as much as 20%, but the high resolutions cannot be achieved.

The gamma-ray spectrum can be measured in a beta-spectrometer by analyzing the electrons ejected by gamma-rays from a thin radiator placed at the source position. The gamma-ray source is placed at a distance behind the radiator depending on the effect used to eject the electrons. The photo-effect produces monoenergetic electrons, which allows the gamma-source to be close to the radiator. Pair production and the Compton effect produce a continuous electron spectrum and in order to narrow the distribution and obtain a line, the gamma source is placed sufficiently far so that a narrow beam hits the radiator. In this case, the spectrometer accepts only electrons ejected forward and their energy spread may be of the order of 1%. The Compton effect and pair production, in order to be used properly, demand special design and specialized spectrometers, while with the photoeffect no special spectrometers are needed. A spectrometer good for internal conversion measurements such as a double focusing spectrometer would be also good for external foto-effect measurements.

2.5. Dispersion in a magnetic field

A spectrometer, quite generally, serves to measure a characteristic magnitude, such as energy, momentum, wave-length or frequency of a given radiation. This implies the ability of apparatus to discriminate between various values of the magnitude which are being measured. In several types of spectrometers, the discrimination is achieved by spacial separation of the radiation beam. An optical spectroscope, for instance, separates a beam of light according to the colors, the separation being achieved by a prism, in which the deviation depends on the wave-length. Once the beam has been separated into its components, their respective deviations can be determined, and then the wave-lengths found by using the physical law governing the deviation of light by the prism.

The process of separating a radiation beam according to a characteristic magnitude is called the dispersion. The use of magnetic spectrometer for measurements of momentum spectra of charged particles is based on the ability of magnetic field to produce the dispersion of a charged particle beam. This can be most easily demonstrated by considering a simple two-dimensional case of electron motion in a plane perpendicular to a homogenous magnetic field.

Let us suppose that electrons are emitted from the sources in the direction of y-axis. If the homogenous magnetic field is perpendicular to the xy plane, the electrons will desribe circles in the xy plane, with radii given by the relation (2.9). Two electrons, having momenta p and $p + \Delta p$, although emitted in the same direction, will separate, describing circles with radii ρ and $\rho + \Delta\rho$, respectively (Fig.2.7). After describing semicircles electrons cross the x-axis at points which are separated by a distance Δx. Quantitatively, the dispersion is generally defined by the derivative of the separation with respect to the magnitude which is measured. In our case, the dispersion γ may be defined as ratio of the separation distance Δx, by the difference of corresponding momenta Δp

$$\gamma = \frac{\Delta x}{\Delta p} \qquad (2.15)$$

In actual spectromer the separation distance is usually measured along the focal surface. Another way to define the dispersion is to relate it to the momentum which is being measured, and the expression has then the following form

$$D = \frac{\Delta x}{\Delta p} \, p = \frac{\Delta x}{\Delta(B\rho)} \, (B\rho) \qquad (2.16a)$$

In this course the dispersion defined by (2.16) shall be used, or D_r defined as

$$D_r = \frac{\Delta r}{r} \cdot \frac{(B\rho)}{\Delta(B\rho)} \qquad (2.16b)$$

2.6. Focusing in a Semicircular Spectrometer

When discussing the dispersion of a charged particle beam by a magnetic field, it was assumed that the beam is parallel to the y-axis. In practice, this would represent a very serious limitation. A radioactive source emits beta particles isotropically in all directions. If only narrow quasi-parallel pencil of β-rays was accepted by spectrometer for analysis, the efficiency would be very low, since such pencils correspond to solid angles of the order of $10^{-5} \times 4\pi$. Obviously it is desirable to use a divergent beam, with an aperture angle as large as possible. Much time is then saved, weaker sources can be measured and lines are obtained in the spectrum, which otherwise would be impossible to observe.

Fig.2.8 shows the trajectories of two electrons in the plane perpendicular to the homogenous field, emitted at small angles $+\gamma$ and $-\gamma$

with respect to y-axis. The two trajectories converge to a focus after 180°. Electrons with higher momenta, leaving in the same directions converge at another more distant point. A homogenous magnetic field can therefore focus a beam diverging in a plane perpendicular to the field direction. Moreover, there is a focal line along which electrons of different momenta are focused.

The optimum postition for the detector is in the focal plane. A spectrometer of this type in which the electrons are moving in and near the plane perpendicular to the homogenous magnetic field, and the detectors are placed at 180°, is called semi-circular spectrometer. It is the simplest and oldest of all types of beta-spectrometers.

The focus of a spectrometer always has a finite extention, and in order to see it, more than two trajectories must be drawn. One finds then that only electrons emitted in the direction of the y-axis desribe exactly 180°, while those emitted at certain angles describe more or less than 180°, depending on whether they are outside or inside the semicircle, and cross the focal plane at points nearer to the source. In Fig.2.9, three characteristic trajectories are drawn, two of them passing near the extremities of the aperture, and the third through its centre. When the aperture is placed symmetrically with respect to the central ray, the two extreme trajectories meet in the same points, and all the other electrons of the same momentum accepted by the aperture cross the focal plane between the points determined by characteristic rays. The extension of the focus for the two-dimensional case, is therefore given by the distance, dx, between the points where characteristic rays arrive at the focal plane. It is easily seen from properties of the triangles inscribed in circles, that the extension of the two-dimensional focus dx, is given by

$$dx = 2\rho - 2\rho \cos\gamma = 2\rho (1 - \cos\gamma) \qquad (2.17)$$

Since γ is small, $\cos\gamma$ can be expanded in a series, giving

$$dx = 2\rho \left(\frac{\gamma^2}{2} - \frac{\gamma^4}{4!} + \frac{\gamma^6}{6!} - \cdots\right) \approx \rho\gamma^2 \qquad (2.18)$$

Taking for γ a typical value of $\gamma = 6^\circ = 0.1$ rad, the width of the focus becomes

$$dx = 0.01\rho$$

In this case, first order focusing is achieved as the extension of the focus is given by small quantities of the second and higher orders. In general, the expression for the width of the focus contains members of small quantities to various orders. For second order focusing, the width should be determined by small quantities beginning with the third order, and so on.

So far, we have considered the ideal case of point source and rays confined to a plane. In reality, the source always has finite dimensions and the electron beam is three-dimensional, which will make further contributions to the width of the focus.

An electron leaving at an angle to the plane perpendicular to the field describes an helix. This can be seen by projecting the electron velocity vector to the plane perpendicular to field lines and in the direction of field lines, Fig.2.10. The magnetic field does not affect the velocity component parallel to it, so that electron motion has a constant component parallel to the field line. Since the other component produces circular motion, the two components together produce a helical motion. There is no focusing in the direction of the field. In this type of spectrometer only one-dimensional focusing is achieved. We shall see later that with special field forms the focusing in the other (axial) direction can be achieved and such types of spectrometers are called double-focusing.

An electron which starts describing a helix falls after 180° at a point (z, x) which has x-component smaller than an electron of the same momentum emitted within y-z plane. This means that an axial opening contributes to the width of the focus.

It is easy to see that the dimensions of the source, the width and the length, will contribute to the width of the focus. This is illustrated in Fig.2.11.

The total instrumental width of the image consists of contributions from the source dimensions, radial and axial openings. This is called instrumental width because it is caused by the properties of the spectrometer itself. Another contribution to the total width comes from physical properties of the source, such as the thickness and the inherent line width which are discussed in the next section.

The total width of the image in the direction of dispersion, which is x-axis in our case, determines the resolution.

A semi-circular spectrometer can be either of constant radius or constant field type, (Fig.2.12).The constant radius type has the detector placed at a fixed position and electrons of different momenta are focused at the detector slit by varying the magnetic field. A semi-circular spectrometer has a long focal plane, along which a photographic plate or an aray of detectors can be placed. A part of the spectrum is then detected simultaneously. It is this property of having a relatively extended focal line which permits the semi--circular spectrometer to be also used as a constant field, variable radius instrument. The semi-circular spectrometer, along with some others which we shall consider later, can be classified as flat spectrometers, because their radial dimensions are usually much larger than axial.

2.7. Line Width in a Semi-circular Spectrometer

The line width and the counting rate are complementary magnitudes in beta spectroscopy. A given line width needed to resolve closely lying internal conversion lines, can often be achieved only at the price of a serious reduction of the counting rate. It always pays to consider various contributions to the line width and try to find the optimum parameters for a given case. We shall present now an elementary discussion of line widths.

The contributions to natural line width come from the widths of nuclear excited state and the width of electron shell in which the transition was converted. The width of a level ΔE is connected with the half-life of the excited state τ by Heisenberg's uncertainty relation

$$\tau \Delta E = h \qquad\qquad\qquad (2.19)$$

A level width $\Delta E = 1$ eV corresponds to a half-life of $\tau = 6.6 \times 10^{-16}$ sec. As internal conversion is important for lower energy transitions, most of the half-lives of the excited states of the nucleus are larger than 10^{-13} sec producing a totally negligible width of less than 10^{-3} eV.

The width of atomic levels from which electrons are expelled are generally larger and increase with the charge of the nucleus. Fig. 2.13 shows the approximate values of level widths of K shell (5). The width of K level is approximately 90 eV at Z = 94 and decreases to about 20 eV at Z = 60. The widths of L subshells are 5 – 10 times smaller. On the other hand it is theoretically predicted that the width of M shell would be larger than for L shell.

In high resolution spectroscopy, especially when low energy transitions are measured, the atomic level widths are not negligible. A 200 KeV transition in uranium, for instance, would have a relative K-line width $dB\rho /B\rho \approx 0.06\%$.

The next contribution to the line width may come from the energy loss of electrons leaving the source. A monokinetic electron beam will after traversing a foil of thickness, dx, have a width at half maximum ΔE given by the Landau formula

$$\Delta E = 0.61 \ \rho \ (Z/A) \ dx/ \beta^2 \ \text{MeV} \tag{2.20}$$

Since electrons are generated in the whole source one can take as the effective thickness approximately one-half of the source thickness dx/2. Putting $Z/A \approx 0.4$, formula (2.20) reduces to

$$\Delta E \approx (0.12/\beta^2) \cdot \rho \ dx \ \text{MeV} \tag{2.21}$$

At the electron energy of 30 KeV, the squared velocity term becomes 0.1 and for a source thickness of 20 $\mu g/cm^2$, the half width is about 20 eV. As electron energy increases to 100 KeV, the width becomes three times smaller.

We can now proceed to instrumental width, which for a given type of spectrometer depends on the angular spertures of the beam, source size and the width of slit in the focus, defining the portion of the beam which can reach the detector.

Let us suppose that the measurements are made with a constant radius semi-circular spectrometer. An ideal source with negligible inherent and energy loss widths would have in the focal plane an image whose width would have been equal to the instrumental with ΔE_i. Let us also suppose, for reasons of simplicity that the density of electrons over the whole image is uniform, which is practically never the case.

At a field setting smaller than necessary to focus the electrons of a given line, the image is falling beyond the detector slit and as the field is increased, the radius of the image is contracting until at a given value a part of the focused beam starts entering the detector (Fig.2.14). The counting rate then starts to increase with the field reaching the maximum when the image coincides with the slit. The maximum is sharp if the slit has exactly the same width as the image. As the field is increased further, the counting rate drops gradually to the continuous background value. In this hypothetical case the line would be symmetrical with a half-width equal to one-half of the extrapolated base width.

Next, let us take into account the finite width of the atomic level ΔE_a, supposing again for simplicity that the distribution is rectangular. The width of the image would increase to $\Delta E_i + \Delta E_a$. The former slit corresponding to the width of ΔE_i would give a line with a flat top and both the half width and the base width would be larger by ΔE_a.

A real source always has some energy absorption and the electron distribution is of the form shown in Fig.2.14b. The image is extended to lower energies with falling intensity. This produces the line shape with a characteristic low energy tail. It should be added that for the same source the tail is energy dependent, being more important at low energies, decreasing, but never disappearing at high energies.

The picture of the width composition we have presented is oversimplified, especially because we have been using the unrealistic, simple rectulangular distributions and supposed that various widths can be simply added:

2.8. "Anatomy" of a Spectrometer

The main components of a spectrometer are

- magnet, with magnetic field measuring system and the earth's magnetic field compensating coils, if necessary

- vacuum chamber containing various beam defining baffles

- source

- detector

- data handling system

We shall desribe them very briefly.

<u>Magnet</u>. - The most important part of a spectrometer is the mag-
net, which has to fulfill the following requirements:

a) Produce a magnetic field of desired geometry within the space
through which the electron beam is passing. It often happens that the field
geometry is correct along the central ray of the beam but starts deviating in
one or two dimensions as the perifery of the beam is approached. In spectrometers with complex field geometries the transmission may be limited by the
relative dimensions of the space containing the correct field. The quality of a
magnet design is then judged by the relative extent of the correct field space.

b) During the measurements at a given setting, the field strength
has to remain constant within limits assigned according to the desired precision. In high precision experiments it is often required to have the field
constant within 0,005%.

c) The geometry of the magnetic field should be independent of the
field strength and remain constant within the whole energy range.

d) Provision has to be made to change continuously the field
strength within wide limits, depending on the electron energy range to be
measured. This energy range is usually from a few keV to 2 - 3 MeV.

A magnetic field of a given geometry can be achieved with three
different categories of magnets.

1) Iron-free magnets. - The magnetic field is produced by combinations of coils, without ferromagnetic materials. As we shall see in the case
of double-focusing spectrometers, a given field geometry can be obtained by
different coil geometries.

The advantages of iron-free magnets are:

- The field shape does not depend on the field strength and remains
constant over the whole range, provided the temperature of the coils is kept
constant, especially at high fields.

- The field is proportional to current, so that the field measurement is reduced to the measurement of coil current, which simplifies the operation.

The disadvantages are:

- Higher power consumption.

- Sensitivity to stray magnetic fields. The iron-free spectrometers are housed in special iron-free buildings, isolated from other laboratories and roads. This increases the cost and diminishes the usefullness, since they cannot be placed near accelerators and electromagnetic separators.

- Earth's magnetic field has to be compensated. This is done with a combination of several pairs of Helmholtz coils. The importance of compensation increases with the size of the spectrometer. For precise compensation it is necessary to monitor the earth's magnetic field and arrange that its time variations are followed by compensating field.

We shall illustrate the three magnet types using the simple case of a semicircular spectrometer with a homogenous magnetic field. Fig.1.15 shows an iron-free semicircular spectrometer described by Arnoult (6). It is a classical pair of Helmholtz coils with the mean radius of the coils being equal to the mean coil distance. The homogeneity of the field is better than 0.1% in the median plane within a distance seven times smaller than the radius of the coils, and along the axis within 1/4 of the coils separation distance. The maximum radius of electron paths is about 1/4 of the coil radius, which shows how large coils are needed in this case.

2) Electro-magnets with iron pole-pieces. - Electromagnets have many applications in electrical engineering and elements of their design may be found in any textbook. Possibly this is the reason why the papers on beta spectrometers almost never contain a description of magnet design, which is unfortunate because the beta-spectrometer magnets represent a class in themselves, with their own characteristics. An analysis of correlations between the design criteria and the actual behavior and performance of the magnet would be very useful.

The difficulties with electro -magnets stem from the nature of magnetic materials themselves. The field is not strictly proportional to the current in the energizing coils because of remanence and non-linearity of the magnetization curve. If there were any local saturations, the field geometry would be affected at high fields. On the other hand, at very low fields, the field geometry may be seriously distorted because one of the critical design rules might cease to be valid. When designing the shape of a pole piece, which should produce a given field geometry it is assumed that magnetic field vector is perpendicular to the surface of the pole piece. This is valid for high permeability, but not at low permeabilities which may be obtained at very low fields.

The fringing fields, which are easier to measure than to predict, limit the useful volume in the gap between the pole-pieces. Their effect depends on the shape of pole-pieces, but usually they affect the field shape at a distance from edges equal to 1 or 1.5 times the separation of pole pieces.

Finally, much depends on the quality of the iron delivered by the producer. It should be homogenous and have the magnetic characteristics which are used in the design.

Because the field is not proportional to the current, it is necessary to have a field measuring device which controls the current.

The advantages of electro-magnets are the following:

- The power consumption is low.

- They are not sensitive to the earth's magnetic field, so that there is no need for large compensating coils.

- They are not very sensitive to stray fields and can be placed in the laboratory close to other apparatus. They are compact and can be moved easier.

- For spectrometers of asymmetrical type such as sectors, the electromagnet are more convenient.

An electromagnet designed by Geoffrion and Giroux (7) for a semi-circular spectrometer is shown in Fig.2.16. Only 24 watts are needed to focus electrons of 10 MeV. Compared to Arnoult's iron-free magnet, the power consumption is about three orders of magnitude smaller, for the same electron energy.

The magnetic field is measured by a comparison with the field of an auxiliary pair of Helmholz coils. Two identical search coils rotate, one in the gap of the electromagnet, the other in the Helmholz field and the difference of induced EMF over a given ratio controls the current in the magnet.

3) Permanent magnets. – So far the use of permanent magnets has been limited to semi-circular spectrographs, which have a long focal plane, allowing simultaneous detection of a substantial part of the spectrum. In most cases, the detection is made by photographic methods.

If a permanent magnet is kept at a constant temperature, its field remains very stable. When its field is changed to another value, however, the fields begin drifting slowly back towards the former value for some time afterwards. This back-drifting may be reduced by recycling, but another convenient solution is to have three or four magnets at different field strengths, covering the whole energy range. Since they would be permanently kept at a given field, their design can be optimized, and there would be no need for field changing coils.

Permanent magnets are suitable only for the constant field variable trajectory type of spectrographs.

Fig. 2.17 shows a permanent magnet spectrograph (8). A 74-cm long photographic plate detects a part of spectrum with momentum range varying from $(B \rho_0)$ to $(6 \times B \rho_0)$.

Alnico V permanent magnets are distributed between the pole-pieces and the yoke. The field in this spectrograph is measured by a simple Cotton balance.

Vacuum Chamber. – The focusing of the beam takes place in the vacuum chamber. Its shape often follows the symmetry of the field or pole-pieces. In lens spectrometers, the chamber is cylindrical, placed coaxially inside the coils, while in flat spectrometers it has often the same shape as pole-pieces. To prevent the scattering of electrons by air, a vacuum greater than 10^{-4} mm Hg should be maintained in the chamber (Fig. 2.18).

The walls of the chamber and all the devices that it contains, such as diaphragms, absorbers and measuring units are exposed to a flux of electrons and photons. Scattered and secondary radiation, if allowed to reach the detector, might seriously increase the background counting rate, which is undesirable. This is reduced in several ways. First, the chamber and the diaphragms are either made of, or lined with a low Z material. On the straight line between the source and the detector, there should be enough material to absorb most of the direct gamma-ray beam. Several anti-scattering diaphragms are placed along the chamber to prevent electrons leaving the walls to reach the counter. The openings of the anti-scattering diaphragms are just large enough to let the beam pass at the maximum solid angle.

When assembling the spectroscope it is important to have the vacuum chamber well aligned along the axis of symmetry. In a lens spectrometer, for instance, the source and the detector should be on the lens axis, and the critical beam baffles should secure the cylindrical symmetry of the beam with respect to the axis of the lens. The critical baffles are the entrance and exit baffles.

The entrance baffle defines the shape of the beam accepted into the spectrometer. In earlier spectrometers the entrance slits had simple forms; they were ring shaped in lenses, and usually rectangular in flat spectrometers. When more was learned about the aberrations and line broadenings due to field imperfections, the shapes of entrance slits became more complex, in order to minimise the effect of aberrations and field imperfections. The usual procedure now is to obtain the iso-aberration curves, which show the required entrance slit controur for a given resolution.

An exit diaphragm is usually placed at focus to define the part of the beam transmitted to the detector. The shape of the exit slit follows the shape of the image. Since usually the width of the exit slit contributes to the line width and the transmission, optimum ratio has to be found between the peak counting rate and the acceptable line width.

Source - Much of the essential information about sources used in beta spectroscopy has already been given earlier in this chapter. The shape and size of the source depends on the focusing properties of the spectrometer. A larger spectrometer allows the use of larger sources. The shape is circular in lenses and mostly rectangular in flat spectrometers.

The source is usually mounted on a rod passing through the vacuum valve, which allows the sources to be exchanged without affecting appreciably the vacuum in the chamber.

The whole source-handling mechanism has to be precise enough to insure the reproducible positioning of the source.

Detector - The detector should satisfy the following requirements, when used with constant trajectories spectrometers:

1. It's shape and size should be such that all electrons passing the exit slit can reach the detection sensitive volume.

2. The detection efficiency should not depend on the energy. It is desirable that the detection efficiency is close to 100%.

3. The background counting rate should be as small as possible.

4. The detection system should not be sensitive to the magnetic fiels.

The detectors used so far with beta-spectrometers are: Geiger counters, proportional counters, scintillation, and semiconductor counters, and electron multipliers.

The Geiger counter is the simplest of them all and by far the most widely used detector in magnetic beta-ray spectrometers. It can satisfy all the above mentioned requirements provided care is taken to have a window thin enough to transmit nearly 100% of lowest energy electrons which are to be measured. The absorption of electrons in a Geiger counter window is illustrated by curves on Fig.2.19 showing the percent of transmitted electrons as a function of window thickness and electron energy (10). As can be seen, a window of 400 $\mu g/cm^2$ would transmit 80% of electrons having an energy of 25 keV and the transmission would fall to 20% for 15 keV. At energies between 10 to 20 keV the windows have to be 30-100 $\mu g/cm^2$ thick. The windows of this thickness cannot withstand the difference between the atmospheric pressure and the pressure of the filling gas, which may be about 10 cm Hg, so that they have to be filled in the spectrometer itself. The vacuum chamber and the detector are evaccuated together and then the detector is filled with an appropriate mixture. Sometimes thin foils have very small holes through which the gas diffuses at a slow rate. To reduce the effect of diffusion, the counter either remains connected with a large volume of filling mixture or gas flows permanently at a constant pressure.

To reduce the background, the counter is enclosed in a lead shield and the active detection volume is not made larger than necessary for proper functioning.

The use of proportional counters meets the same window problems but offers the possibility of background reduction and faster counting.

When a beta-spectrometer is coupled to another spectrometer for coincidence measurements, much faster detectors are employed, such as organic scintillators: antracen, stilben or plastic mixtures. Then one has to take care of the photomultiplier's sensitivity to the magnetic field, Three different solutions are available: first, field compensation; second, magnetic shielding of the photo multiplier with μ -metal; or third, placing a light guide between the detector and the photo multiplier removed to a field-free region. Sometimes two or all three methods are used in the same spectrometer.

In magnetic spectrographs having an extended focal plane, for many years the only detector used was the photographic plate. Since the relation between the total intensity of electrons hitting a small surface of the plate and its blackening is not linear but is given by a curve of saturation type, the photographic plate is not convenient for precise measurements of intensities. This defficiency is not present in electron sensitive nuclear emulsions, where individual electron tracks are counted. These are also sensitive to much smaller intensities, but their operation is much more complicated. Because of great radiation sensitivity of electron emulsions, one must either use fresh plates or prepare them from liquid emulsion and then one faces the task of counting electron tracks.

Recently, arrays of semiconductors have begun to be used. One of their main advantages is the possibility of applying the multichannel data handling techniques. This represents an important advance, which will probably stimulate a greater use of extended focal plane spectrographs in the future.

When a large background is present, it can be reduced by using more than one counter in coincidence. The simplest way is to have two Geiger counters in the focus, one after another, counting only the particles which pass through both of them (11). If there is an intermediate focus, a Geiger counter can be placed in it and have it in coincidence with the main detector. Care has to be taken about scattering in the intermediate detector.

The detection of low energy electrons, of the order of keV, requires special arrangements. One possibility is to accelerate electrons, raising their energy to values sufficient for passing through the window. The electric field device can be placed either at the source or at the detector end, and the accelerating potentials are usually below 10 keV. The preacceleration is inconvenient because the relative separation of close lying lines decreases. The post-acceleration requires care, not to produce any discharge (12).

Recently low energy electron detectors were developed based on electron multiplication in windowless tubes and are commercially available.

Power supplies - The primary concern in the choice of power supply is to obtain a stable magnetic field. Stabilization of one part in 10^6 can be achieved.

Computer control of spectrometers - Computers are beginning to be used with beta spectrometers. A computer program may include instructions on measurements, such as time or number of counts for each point, the distance between points and the parts of the spectrum to be covered. Once the spectrum is obtained, the computer can then find the energies and intensities of the lines.

2.9. Performance Parameters

The performances of a spectrometer are characterized by several quantities, the most important being: dispersion, resolution, transimission and luminosity. Let us consider them in turn.

Dispersion measures the relative separation between the images of two monokinetic groups of electrons. As was shown in 2.5 the dispersion can be defined as

$$Y = \frac{ds}{dp} \qquad D = \frac{ds}{dp}\, p \qquad D_r = \frac{dr}{r} \cdot \frac{p}{dp} \qquad\qquad (2.22)$$

where ds is the distance between the high energy edges of two images of electrons having a momentum differing by dp. Two adjacent lines cannot be separated if their full image width dx is not smaller than ds. A larger dispersion
allows the use of larger sources.

Dispersion is easily measured in those types of spectrometers, in
which the beam is deflected along circular orbits. It then has an expression of
the form

$$D = ar$$

where r is radius of the mean orbit (optical circle), and a is a constant
depending on the field form. Dispersion may be increased, either by increasing
the dimensions of spectrometer, or by chosing the field form giving a higher
value of the coefficient a.

Resolution or Spread is the relative half-maximum width of a
monokinetic electron line, quantitatively defined by

$$R = \frac{dp}{p} = \frac{d(B\rho)}{(B\rho)} \tag{2.23}$$

where $B\rho$ is the momentum of the line and $d(B\rho)$ is the width of the line at
half maximum.

The resolution can be connected with dispersion by writing D in
(2.22) as

$$dp = \frac{ds}{D} p$$

and substituting in (2.23), to obtain

$$R = \frac{ds}{D} \tag{2.24}$$

where now ds is half of the image width. Relation (2.24) shows that a larger
dispersion gives a better resolution.

In flat spectroscopes where the beam envelops a circular equilibrium orbit of radius r_o, the resolution has usually the following general
form

$$R = (1/2) \; \Sigma_i \, \omega_i \cdot 1/(b \, r_o) \tag{2.25}$$

where $\Sigma_i \omega_i$ is the sum of partial base widths and b is a coefficient depending on the magnetic field form.

In theoretical calculations it is sometimes easier to operate with base width R_o and suppose that $R = R_o/2$.

The resolution can be improved by:

1. Reducing source dimensions

2. Reducing the apperture angles

3. Reducing the exit slit width

4. Increasing the radius of equilibrium orbit r_o

5. Eliminating aberration terms to a higher order

6. Chosing a field shape giving a larger coefficient b.

The concept of resolving power is less used in beta spectroscopy. The resolving power η is equal to the reciprocal value of resolution

$$\eta = 1/R \qquad\qquad (2.26)$$

Solid Angle Ω , expressed in percentage of 4π , desribes what part of isotropically emitted radiation from the source is accepted by the entrance baffle.

A part of the beam accepted by the entrance baffle can be lost before reaching the exit slit. In lens spectrometers, for instance, the supports of central absorbers and baffles cut a relatively smaller part of the accepted beam. Serious field imperfections may cause a partial loss of the beam. A beam loss appears unexpectedly sometimes, at relatively large solid angles, when the field has not been sufficiently investigated or approximations used in the design are inadequate. Whenever a newly designed spectrometer is investigated it is wise to test carefully whether all electrons accepted by the entrance baffle reach the detector slit.

The image produced by the accepted beam of monokinetic electrons is often larger than the exit slit. Since the width of the exit slit contributes to the line width a term of first order, whenever resolution is important a compromise has to be made between the peak counting rate, which favors larger slit width, and the line width requiring smaller exit slit width. When the slit is exactly equal to the image all electrons reaching the slit are transmitted to the detector. It is supposed that the line is strictly monokinetic and the field intensity is set to focus these electrons. The counting rate represents then a fraction T of the emission rate. The fraction T expressed in precentage represents the <u>Transmission</u> of the spectrometer. In the ideal case, that there is no beam loss between the source and the detector, so defined transmission is numerically equal to the solid angle.

In practice the counting rate is usually smaller than possible for a given transmission and the reduction can be numerically expressed by the <u>Acceptance Factor</u> F, which is smaller or equal to 1. The product FT represents the Effective Transmission at a given resolution. The values of F are often from 0.5 to 0.8. An illustration of F factor is shown on Fig.2.20. The flat top line was obtained with a slit several times wider than the image, and the ratio of small line height to the large line height represents the F factor.

It should be mentioned that experimental determination of F and T require a line with negligible low energy tail.

The counting does not depend only on the effective transmission, but also on the dimensions of the source and its specific activity. Practically, in most cases it is convenient to be able to use larger sources, since then a total activity can be spread in a thinner layer and low energy tail reduced. For that reason the performance of a spectrometer is better judged by the product of source area and transmission, called <u>Luminosity</u>. Three different definitions of luminosity are possible:

Geometric Luminosity $L_g = A \cdot \Omega$

Luminosity $L = AT$

Effective Luminosity $LF = ATF$

The reader should be warned that there is some terminology confusion about luminosity as well as transmission parameters. Often authors call simply luminosity any of the three parameters defined above. Since it is difficult to impose a uniform terminology, it is important to find first the authors own definition of terms he uses.

2.10. Classification of Spectrometers

The briefest description of a spectrometer (an identification card) should contain the following information:

1. Beam geometry - In many cases, but not always, beam geometry can be simply described. In the lens spectrometers, the beam has an axis of symmetry, while in the flat spectrometers, beam remains close to a median symmetry plane.

2. Field symmetry - Most of the spectrometer types are based on cylindrically symmetric fields which are simpler for theoretical treatment and the magnet design is in principle less laborious than in the case of fields which do not possess cyllindrical or rotational symmetry. Very often there is a symmetry plane perpendicular to symmetry axis.

3. Field geometry - The geometry of the magnet field is defined by describing the field variation in the significant symmetry plane or axis. Thus for flat spectroscopes the field variation in median plane is in most cases sufficient to describe the whole field, while in a lens, it is sufficient to know the field variation along the symmetry axes.

4. Dimensions of focusing - Focusing can be single, as in semi--circular spectroscopes, or double. There are also intermediate cases when the field is focusing in both dimensions, but at different positions. The double focusing usually implies that the focus in one dimension coincides with the focus in the second one.

5. Type of magnet - Three types of magnets are in use: iron-free, iron-core, and a special version of iron-core, which is rarely used, permanent magnet.

6. Focus extension - The terms spectrometer and spectrograph denote two limiting cases. In the first case a single detector, with a single exit slit, records the radiation intensity for each field intensity as it varies stepwise. A spectrograph records a part of a spectrum for a given field intensity and was originally used only for photographic recording. Recent developments of detectors and multichannel data handling techniques stimulated the use of multidetectors. There may be any number of them, from one to hundred, and thus former sharp distinction between a single electrical counter and a long focal plane detector is disappearing. The multidetector systen increases the number of informations simultaneously registered and represents an important indicator of the instruments data acci nulating efficiency.

7. Degree of focusing – The degree of focusing relative to an opening angle is determined by the lowest power of that angle which appears in the expression for the image width. The second power corresponds to first order focusing, the third power to the second order, etc. It should be noted that the first power corresponds to no focusing in that dimension. In the double focusing instruments often the order of focusing is not the same in both dimensions.

8. Size – It is often sufficient to illustrate the size of a spectroscope by a characteristic length such as the radius of the optic circle in the flat type or source–detector distance in a lens.

The classification in this book is based on beam geometry, field symmetry and field geometry. It is not a strict classification, because it was considered more important to put together the cases offering some common elements for theoretical treatment and the possibility to proceed from simpler to more complicated instruments.

References for Ch.2

1. D.Fehrenz and H.Daniel, Nucl. Instr. Meth. 10, 185 (1965)

2. H.Paul, Nucl. Instr. Meth. 37, 109 (1965)

3. J.G.Balfour, J.Sci. Instr. 31, 395 (1954)

4. D.D.Douglas, Phys. Rev. 75, 1960 (1949)

5. G.T.Ewan and R.L.Graham, in α, β , γ - spectroscopy, ed. K.Siegbahn, North - Holland, 1965

6. R.Arnoult, Ann. Phys. 11 sér. 12, 239 (1939)

7. C.Geoffrion and G.Giroux, Can. J.Phys. 34, 920 (1956)

8. M.Mladjenović, Bull. Inst. Nucl.Sci. "B.Kidrič", 6, 53 (1956)

9. Z.Body and D.Berenyi, Acta Phys. Acad. Sci. Hung. 15, 215 (1963)

10. R.O. Lane and D.J. Zaffarano, Phys. Rev. 94, 960 (1954)

11. M.Mladjenović and A.Hedgran, 8, 49 (1954)

12. Pre - and post - acceleration is discussed in following papers:

    L.M.Langer and C.S.Cook, RSI 19, 257 (1948)

    H.M.Agnew and H.L.Anderson, RSI 20, 869 (1940)

    D.K.Butt, Proc. Phys. Soc. London A 63, 986 (1950)

    C.H.Chang and C.S.Cook, Nucleonics 10, 24 (1952)

    R.D.Hill, J.W.Mihelich and M.T.Pigott, RSI 21, 498 (1950)

    C.Geoffrion and G.Giroux, Can. J.Phys. 34, 153 (1956)

    A.Moussa, J.Phys. Rad 19, 94 (1958)

    S.Rosenblum, J.S.Dionisio and M.Valadares, J.Phys. Rad 17, 112 (1956)

    J.Delesalle, J. Phys. Rad. 19, 111 (1958) and J.Res.CNRS 8, No41, 237 (1957)

    H.Slätis, NIM2, 332 (1958) and Ark. Fysik 29, 485 (1965)

    T.Novakov, J.M.Hollander and R.L. Graham, NIM26, 189 (1964)

    W.Mehlhorn and R.G.Albridge, NIM26, 37 (1964)

Text to Figures. Ch.2

Fig. 2.1.   Cartesian coordinate system moving with the particle. The axis are the tangent $\vec{t}$, the normal $\vec{n}$ and the binormal $\vec{b}$ to the curve.

Fig. 2.2.   The relation between T and $B\rho$ for electrons.

Fig. 2.3.   The relation between $d(B\rho)/(B\rho)$ and dT/T, shown as function of energy.

Fig. 2.4.   (a) The influence of the souce thickness on the shape of beta spectrum of ^{32}P, measured by Fehrentz and Daniel (1). The thicknesses of the sources are 0.01, and 10 mg/cm^2.

             (b) The broadening of internal conversion line due to the source thickness. The electron energy is 327 keV (2).

Fig. 2.5.   The efect of source backing on the electron lines is illustrated by the change of the spectrum A into B, when a thick silver plate is added to the thin backing. The difference is shown in (b). (3).

Fig. 2.6.   The source charging chifts four electron lines having energies between 47 to 140 keV. When Al backing is changed by nylon backing, the electron energy decreases.

Fig. 2.7.   Dispersive action of a homogenous magnetic field.

Fig. 2.8.   Focusing action of a homogenous magnetic field.

Fig. 2.9.   The three principal rays forming an extended focus. Two typical cases are shown.

Fig. 2.10.   The helix described by an electron emmitted at an angle to the plane perpendicular to the (homogenous) magnetic field.

Fig. 2.11. (a) A linear source placed along X-axis contributes its own
length to the image width.
(b) A linear source placed along Z-axis contributes to the
width of the image.

Fig. 2.12. (a) Constant radius spectrometer
(b) Constant field spectrometer

Fig. 2.13. The widths of K-level Γ_k and L - levels Γ_L shown as
functions of Z. (5)

Fig. 2.14. (a) The line shape obtained when the electron beam sweeps
over the detector slit.
(b) Total line shape obtained when ading   inherent and absorption
widths   while the exit slit remains constant.

Fig. 2.15. Iron free semi-circular spectrometer. (6)

Fig. 2.16. Iron core semi-circular spectrometer (7)

Fig. 2.17. Permanent magnet spectrograph. (11)

Fig. 2.18. Continuous beta-spectrum of ^{60}Co at different vacuum values.
The path length is 50 cm. (9)

Fig. 2.19. Transmission curves for Geiger counter windows. (10)

Fig. 2.20. Illustration of F factor. The source width is 0.2 mm, and
the small line was measured with an exit slit 0.2 mm wide
which is smaller than the width of the beam. The large flat
top line was obtained with a slit 5 mm wide, which is wider
than the beam.

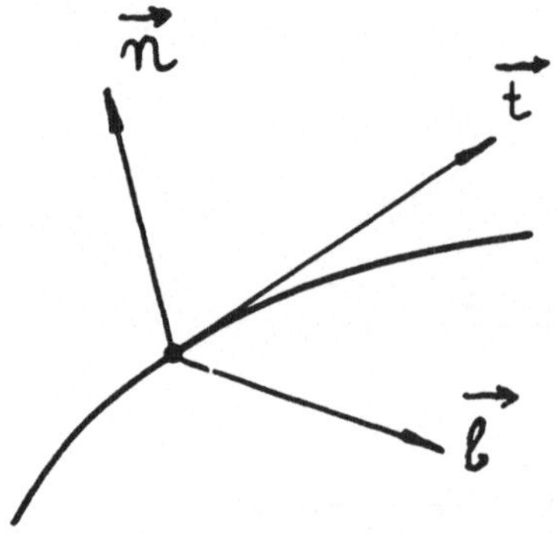

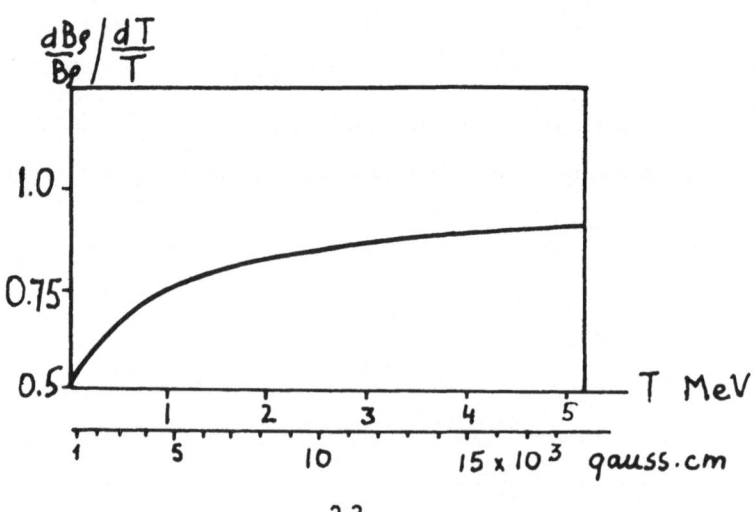

2.1

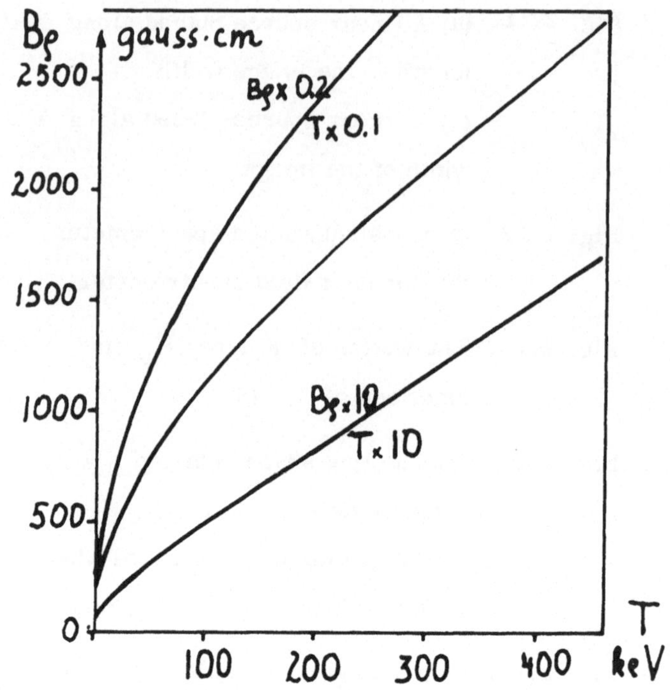

$B\rho$ gauss·cm

$B\rho$ × 0.2
T × 0.1

$B\rho$ × 10
T × 10

T keV

100   200   300   400

2.2

$\dfrac{dB\rho}{B\rho}\Big/\dfrac{dT}{T}$

1.0

0.75

0.5

T MeV

1   2   3   4   5

1   5   10   15 × 10³ gauss·cm

2.3

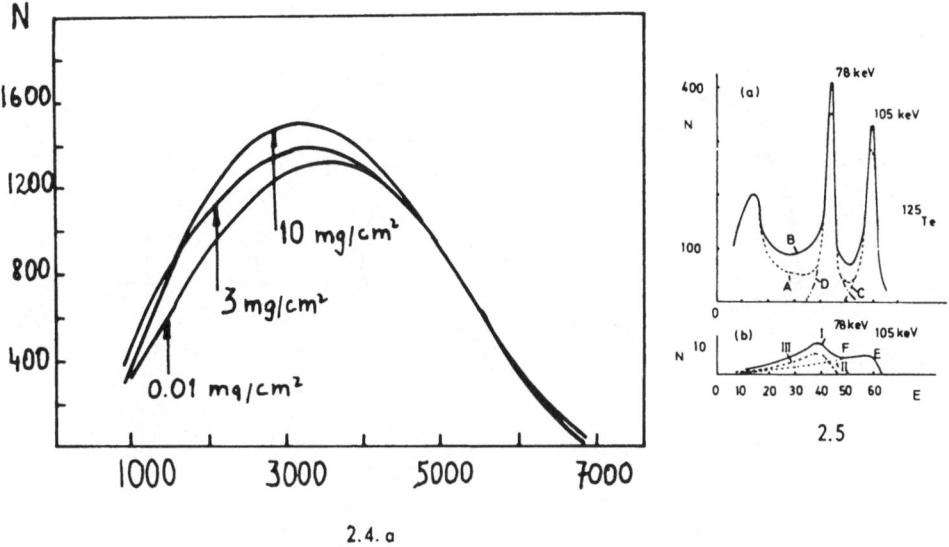

2.4.a

2.5

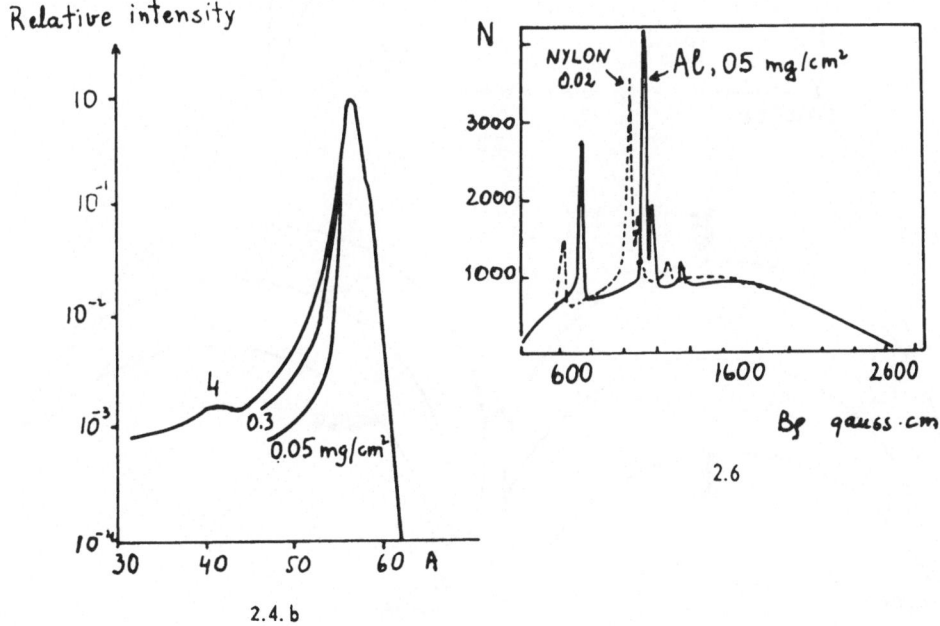

Relative intensity

2.4.b

2.6

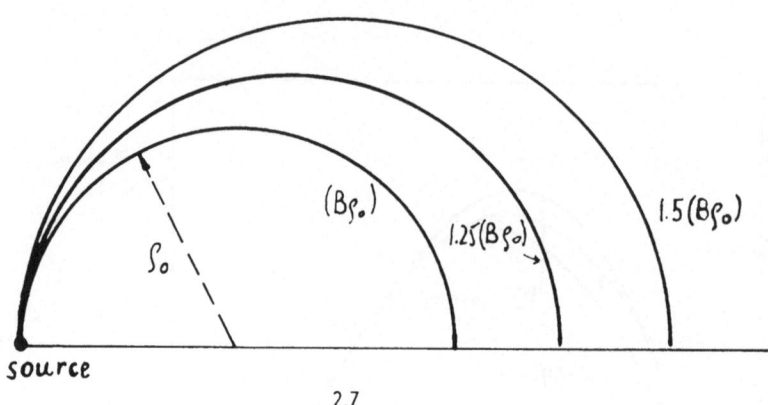

2.7

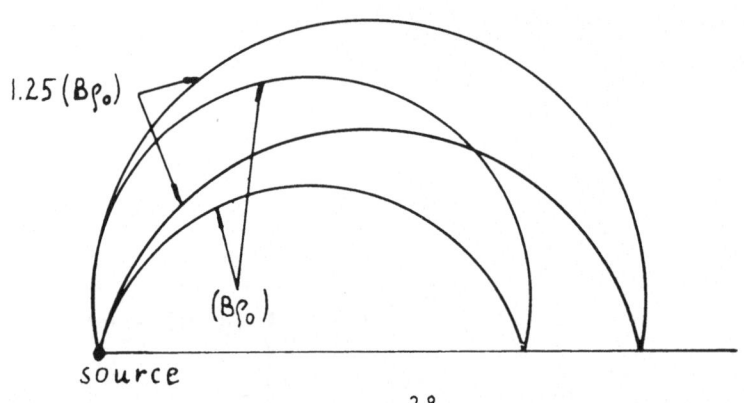

2.8

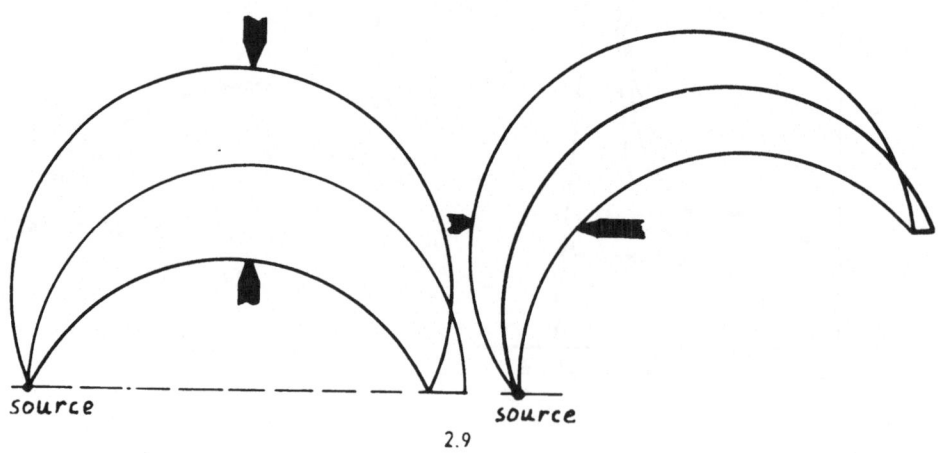

2.9

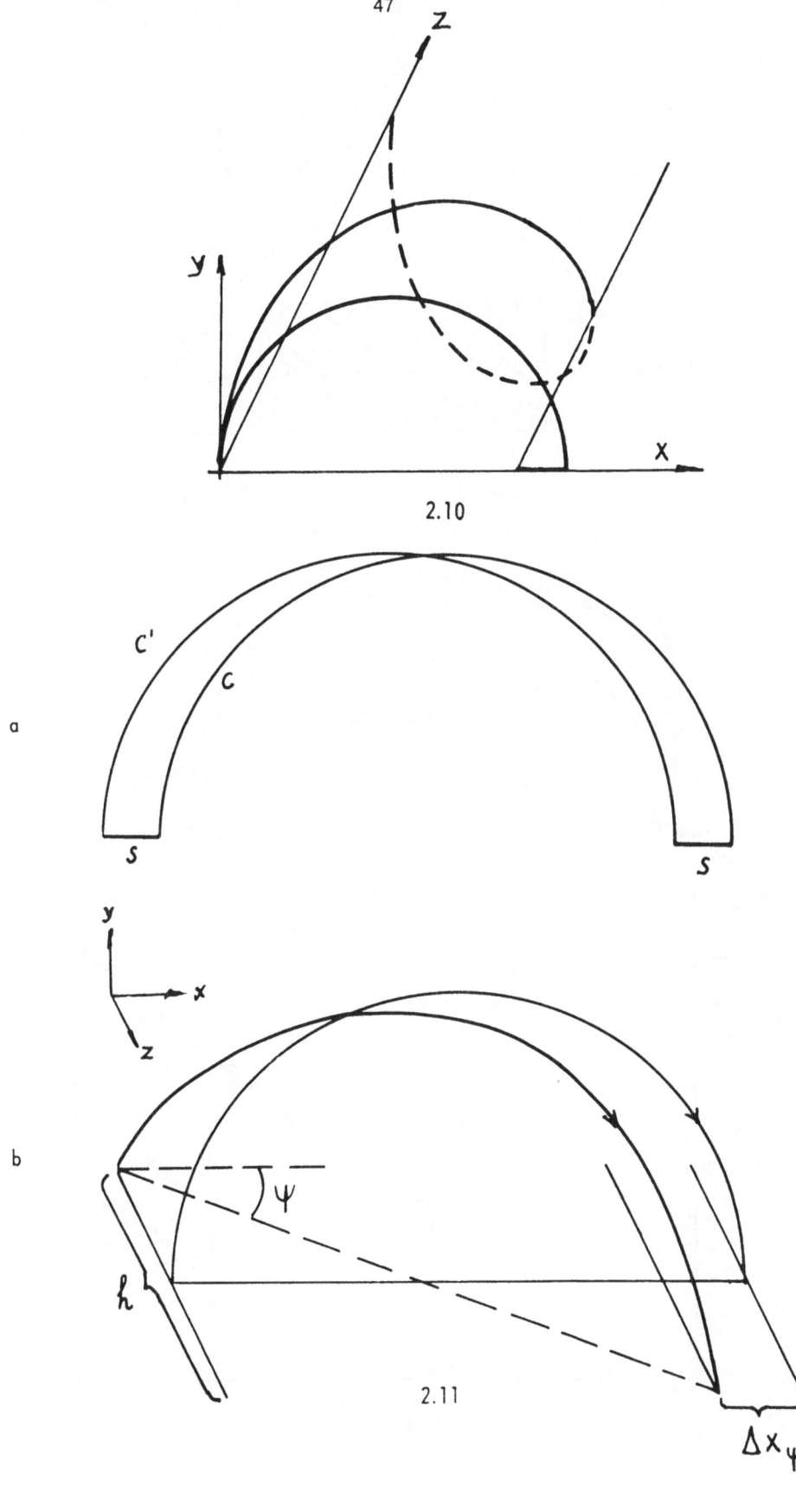

47

2.10

a

C'

C

S   S

2.11

b

ψ

h

Δx_ψ

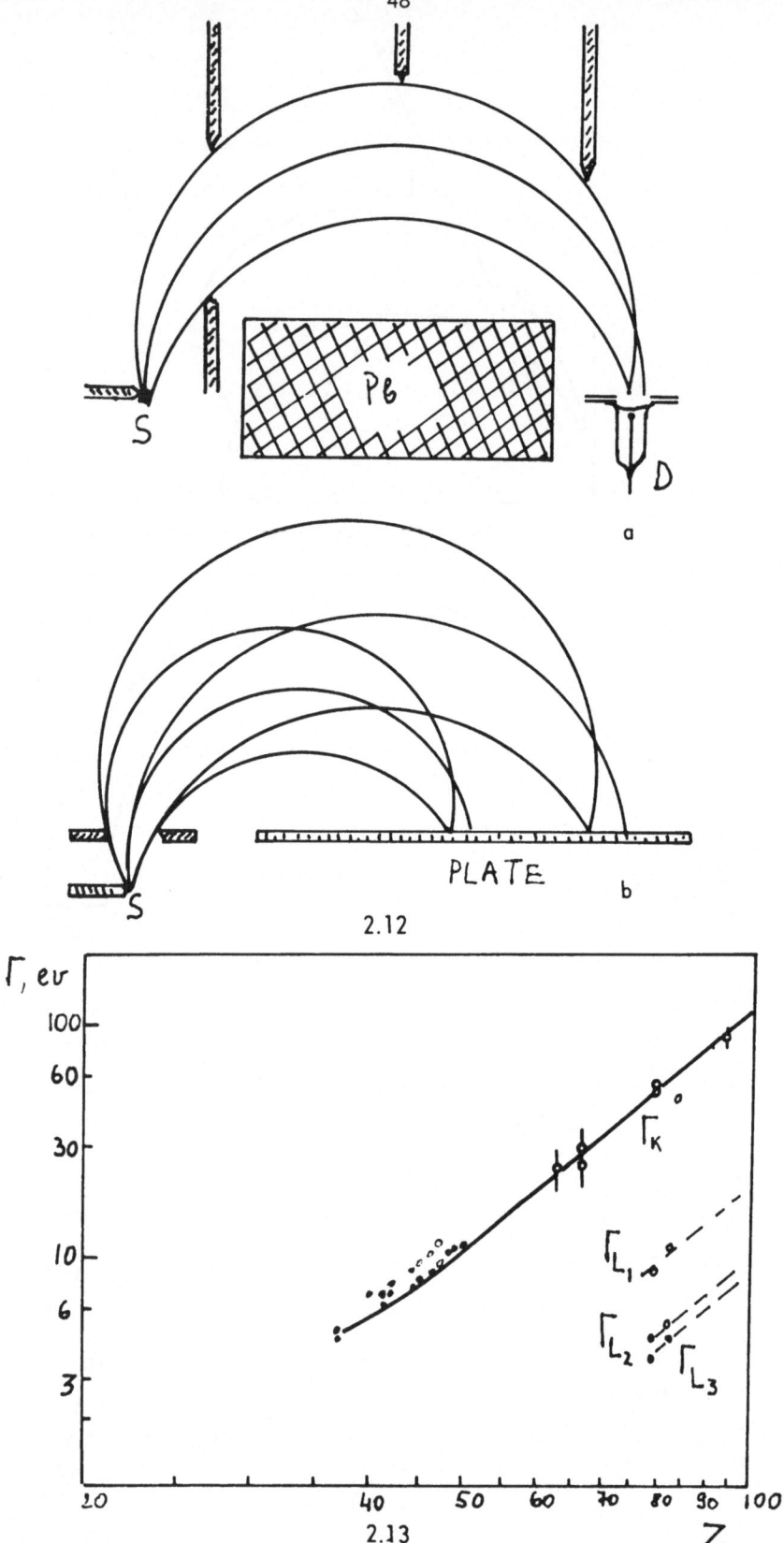

2.12

2.13

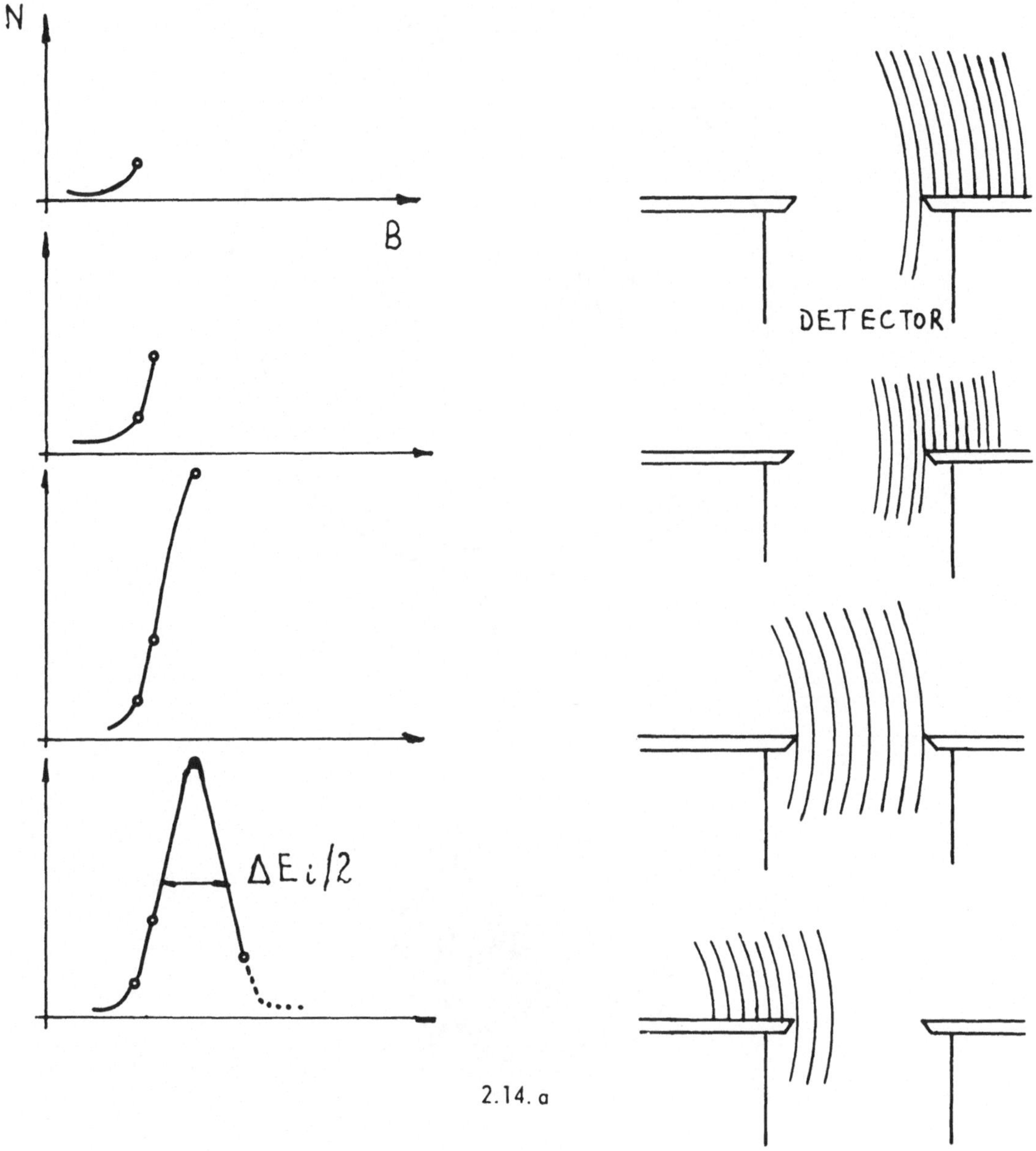

N

B

DETECTOR

$\Delta E_i / 2$

2.14. a

V

$$\frac{\Delta E_i}{2} + \Delta E_a$$

B

2.14. b

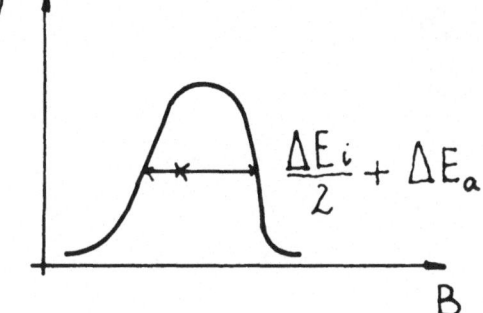

COILS

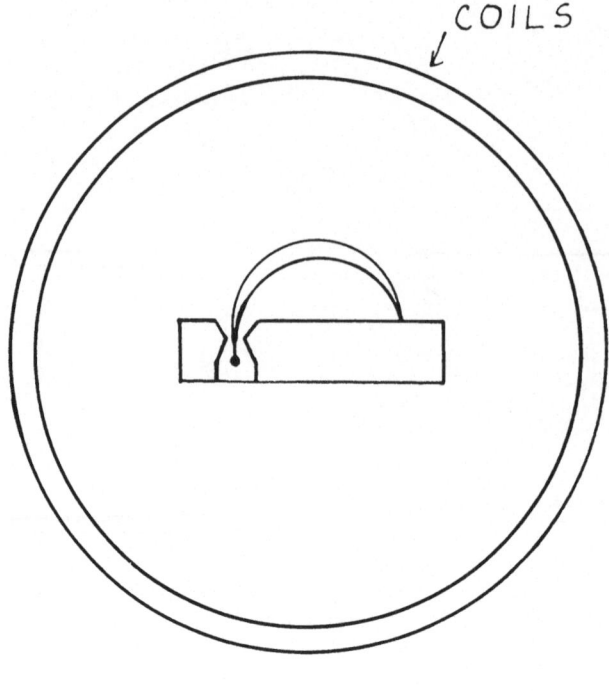

2.15

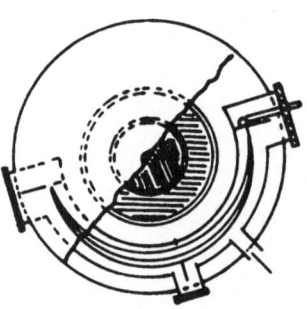

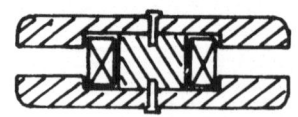

2.16

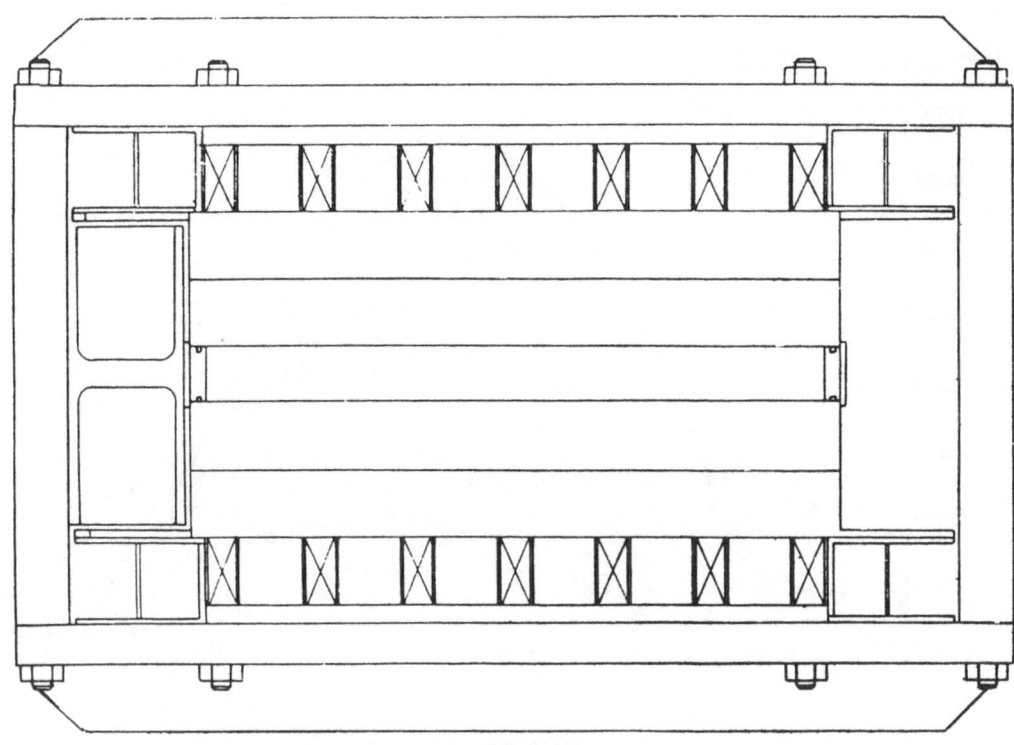

2.17.a

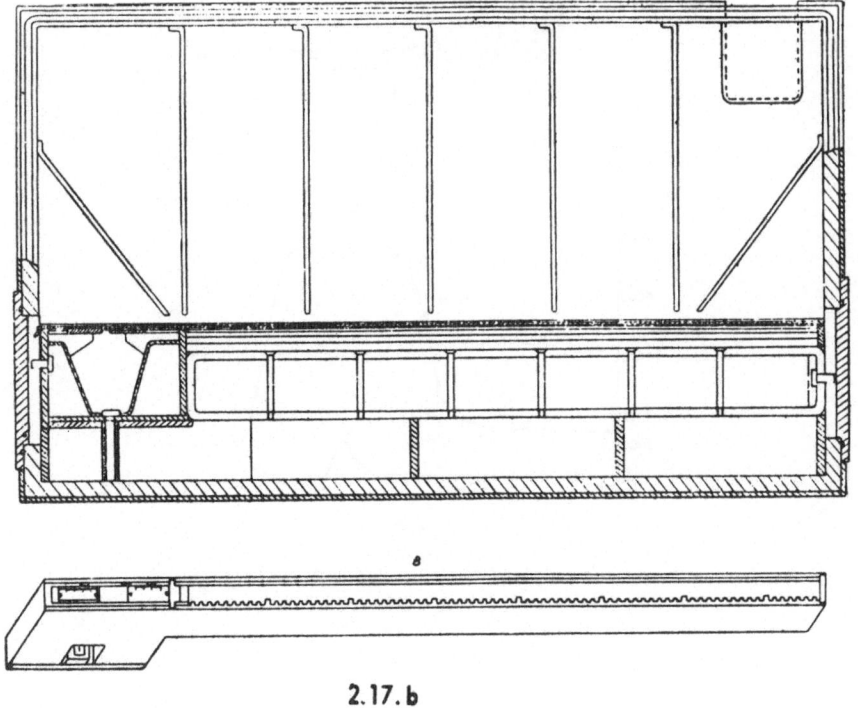

2.17.b

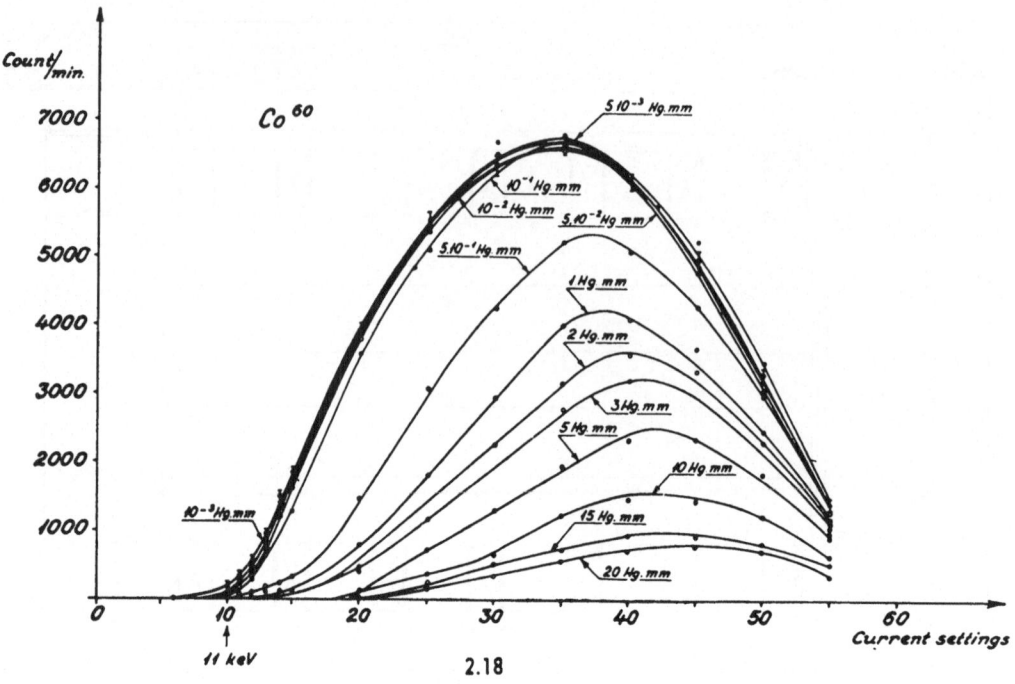

2.18

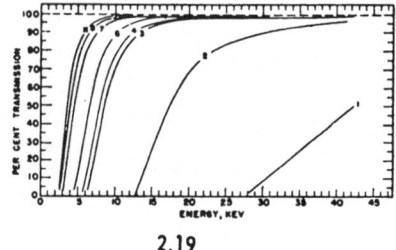

2.19

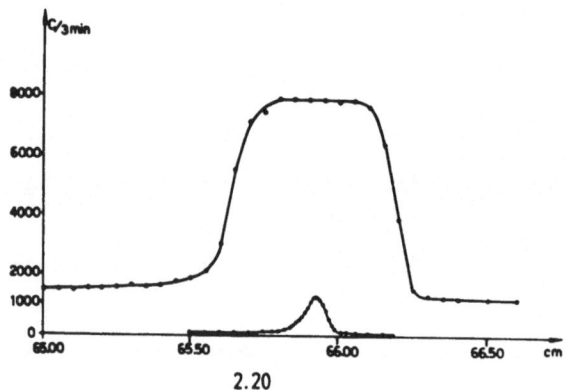

2.20

3. ANALOGY WITH GEOMETRICAL OPTICS

Electro-magnetic fields may act as charged particle prisms, lenses and mirrors. The motion of charged particles can be studied using concepts and methods of geometrical optics. In this chapter the analogy between geometrical and electron optics is briefly considered.

3.1. Fundamental Laws

The geometrical optics are based on Fermat's principle, which can be expressed as

$$\delta \int_A^B \mu \, ds = 0 \tag{3.1}$$

where μ is the index of refraction, and ds is an element of light path. From Fermat's principle follow the first three basic laws of geometrical optics: (1) The rectilinear propagation of light where μ is constant, (2) The law of reflection, and (3) The law of refraction. The theoretical bases of geometrical optics is completed by adding to these laws the assumption that light rays are independent.

In a similar way, electron optics can be based on Maupertuis principle and on the assumption that the interaction between charged particles in a beam may be neglected, which is completely justified in the case of beta-spectrometer. The Maupertuis, principle can be expressed as

$$\delta \int_A^B \left| p - e \, (\vec{A} \cdot \vec{s}) \right| \, ds = 0 \tag{3.2}$$

where A is the magnetic vector potential.

The index of refraction of an electro-magnetic field is therefore given by

$$\mu_{el.m} = p - e \, (\vec{A} \cdot \vec{s}) \tag{3.3}$$

When a magnetic field is absent, the index of refraction reduces to the first term, which is a scalar

$$\mu_{el} = p \qquad\qquad (3.4)$$

The momentum of a charged particle in an electric field varies continually, so that refractive index also varies continually. This is generally not the case in geometrical optics. A more detailed discussion of the refractive index of an electron field can be found in Grivet's article (1).

The refractive index of a magnetic field has several properties fundamentally different from one of classical optics. It depends on the direction of motion, and can be arbitrarily modified by adding to $\vec{A}$ a function of the form grad f(xyz), without changing the path of the particle. For that reason it is seldom used in practical problems.

3.2. Ideal Lens

Geometrical optics deals with light phenomena in which the wave length can be considered to be negligibly small in comparison with the other dimensions of the system. It is then possible to define point sources, light rays, and sharply bounded pencils of light rays as geometrical idealisations, and the optical laws can be described in terms of these geometrical concepts. Geometrical optics define ideal systems, which approximate real system to various degrees.

For our purpose, it is sufficient to consider an ideal lens having a cylindrical symmetry. In electron optics such lenses are produced by different kinds of coils, with or without iron-cladding. Magnetic lenses, especially short iron-clad lenses are used often in electron microscopy, while in beta spectroscopy they are at present less important. As it will be seen in later chapters, the performances of some other types are superior and the only advantage of the lens is its simplicity of construction.

The ideal lens which shall be considered corresponds to a short magnetic lens in which the source and the image are in a field free region. Its properties have been extensively studied and complete accounts can be found in books by Glaser (2) and Picht (3). The concepts introduced to describe the action of the ideal lens can be generalized and applied to other focusing devices.

It should be stressed that the definition of an ideal lens and the laws governing its action are perfectly general and independent of the real physical systems which approximately behave in a similar way. Optical devices can be constructed for focusing all kinds of particles, from photons and charged particles to neutrons.

The action of a lens is often limited to a given region, outside of which the rays are straight lines, as shown on Fig.3.1. It divides the space in two parts, the side containing the object being called object space and the other image space. The active region may be so large to include the object and its image, as is often the case in beta spectroscopy.

An ideal lens may be defined as an imaging device, acting in the following way:

1) All rays starting from a point in object space converge to, or diverge from a point in image space.

2) If the object is a plane surface, perpendicular to the axis of the instrument, the image lies in a plane perpendicular to the axis.

3) The image of an object on this plane must be similar to the object, although linear dimensions may be altered.

The difinition of a lens gives essentially the properties of transformation from object space to image space. The transformation is of point--to-point, line-to-line and plane-to-plane type, which can be mathematically expressed by general formulae of colinear transformation.

$$x^1 = \frac{a_1 x + b_1 y + c_1 z + d_1}{ax + by + cz + d} \tag{3.5a}$$

$$y^1 = \frac{a_2 x + b_2 y + c_2 z + d_2}{ax + by + cz + d} \tag{3.5b}$$

$$z^1 = \frac{a_3 x + b_3 y + c_3 z + d_3}{dx + by + cz + d} \tag{3.5c}$$

Where x, y, z and x^1, y^1, z^1 are coordinates in two separate coordinate systems, the first connected with the object and the second with the image space. The above equations are simplified when the system has a symmetry axis, taken as the $x_1 x^1$ axis and it is taken into account that x^1 must be independent from y_1 and that y^1 only changes in sign with no change in magnitude when y reverses in sign. Equations (3.5) are then reduced to

$$x^1 = \frac{a_1 x + d_1}{ax + d} \tag{3.6a}$$

$$y^1 = \frac{b_2 y}{ax + d} \tag{3.6b}$$

Ray tracing is greatly facilitated by the use of cardinal points: two focal points, two principal points, and two nodal points.

The focal plane is defined as conjugate to a plane at infinity. It is seen from (3.6) that x^1 and y^1 become infinite when

$$ax + d = 0 \tag{3.7}$$

This is the equation of the focal plane in object space. By expressing x, y in terms of x^1, y^1, it can be seen that the equation of the focal plane in image space is

$$ax^1 - a_1 = 0 \tag{3.8}$$

The intersections of these two planes with the symmetry axis define the FOCAL POINTS

Taking the focal points as new centers of the coordinate systems, and introducing two new constants f, f^1 defined by

$$f = \frac{b_2}{a} \quad \text{and} \quad ff^1 = \frac{d_1 a - a_1 d}{a^2} \tag{3.9}$$

equations (3.6) become

$$xx^1 = ff^1 \tag{3.10}$$

$$\frac{y^1}{y} = \frac{f}{x} \tag{3.11}$$

These equations give image point (x^1, y^1) when f and f^1 are known. The physical meaning of f and f^1 may be understood by considering the conjugate planes given by

$$x = f \qquad x^1 = f^1$$

The equation (3.11) gives then

$$\frac{y^1}{y} = 1$$

which means that the transverse linear magnification is unity. These planes are called PRINCIPAL PLANES, and their intersections with symmetry axis PRINCIPAL POINTS. It is seen from Fig. 3.1 that rays parallel to axis and deviating into focus have the virtual intersection between the two parts in the corresponding principal plane, because then the lateral magnification between two conjugate points such as A and B is unity.

Finally, NODAL POINTS are conjugate points on the axis, such that a ray aimed at one nodal point proceeds as if it was starting from the conjugate nodal point and in the same direction. This follows from the definition of nodal points as conjugate points with angular magnification equal to unity. Angular magnification may be expressed as ratio of tangents of two angles that a ray is making with the axis. From the Fig. 3.2 it is seen that angular magnification γ is equal to

$$\gamma = \frac{\tan u^1}{\tan u} = \frac{-f}{x^1} = \frac{-x}{f^1} \qquad (3.12)$$

The nodal points are then quantitatively determined by $\gamma = -1$ which gives

$$x^1 = f$$

and by using (3.12)

$$x = f^1$$

It is seen that when $f = f^1$, as it is the case with magnetic lenses, the nodal points coincide with principal points.

The knowledge of cardinal points permits one to find the conjugate of any point in the space. It is sufficient to draw a ray parallel to axis which passes through the second focus, and another ray passing through the first focus leaving the lens in the direction of the axis. The intersection of these two rays give the conjugate point. It is convenient to introduce the distances from conjugate points to principal planes, s and s^1, which are equal to

$$s = f - x$$

$$s^1 = f^1 - x^1 \qquad\qquad (3.13)$$

Combining (3.12) and (3.13) one obtains

$$\frac{f}{s} + \frac{f^1}{s^1} = 1 \qquad\qquad (3.14)$$

which in case $f = f^1$ becomes

$$\frac{1}{s} + \frac{1}{s^1} = \frac{1}{f} \qquad\qquad (3.15)$$

This is the well known lens formula.

To close this discussion of the ideal lens, it would be useful to derive Lagrange law, which can serve for comparison with physical systems. Equations (3.10 - 3.12) combined give

$$yf^1 \tan u = y^1 f \tan u^1 \qquad\qquad (3.16)$$

which is the Lagrange law.

Thus, starting from the definitions of an ideal lens and its cardinal points, and translating them into mathematical language, the lens formula and the Lagrange law may be easily derived using only algebra and geometry.

3.3. Real Lenses

In a real imaging system the rays propagate according to the principle of least action. A real lens would behave as an ideal lens if the laws such as the Lagrange law are compatible with the principle of least action. This question was first considered in geometrical optics and it was found that the real lenses have to obey the Abbe's sine law. It can be derived either from the principle of conservation of energy or from Fermat's principle, and has the form

$$ny\sin u = n^1 y^1 \sin u^1 \qquad\qquad (3.17)$$

where n, n^1 are diffraction indexes.

Comparing Abbe's sine law with Lagrange law (3.16), it is seen that they cannot be satisfied simultaneously for every value of u. Consequently, a real imaging system can never act as an ideal system. Only approximations are possible.

3.4. Gaussian Systems

An approximation mostly used in optics is Gaussian, in which only rays close to axis are considered. Angles u, u^1 are then very small, so that trigonometrical functions of these angles can be developed in series and terms of second and higher orders neglected. Also neglected are the terms involving squares and higher powers of off-axis distances.

In the Gaussian approximation, the Lagrange law becomes

$$nfu = n^1 f^1 u^1 \tag{3.18}$$

while Abbe's sine law is reduced to

$$nyu = n^1 y^1 u^1 \tag{3.19}$$

They are equivalent, provided

$$\frac{f^1}{f} = \frac{n^1}{n} \tag{3.20}$$

which connect focal lengths with diffraction indexes.

In electron optics, the Gaussian approximation is widely applied for design of electron microscopes, while in beta-spectroscopy its importance is reduced as wider apertures and higher order focusings are desirable.

3.5. Aberrations

The differences between the ideal and the real imaging systems are specified in terms of various aberrations. They can be calculated by taking into account the higher order terms. Thus, in lenses odd terms contribute, so that one has third-order, fifth-order errors and so on. In magnetic lenses there are five third-order aberrations: spherical aberration, distortion, curvature of the field, astigmatism, and coma. To these must be

added another important kind of error which is caused by the deviation of
the magnetic fields from perfect symmetry.

Detailed accounts on aberrations may be found in books on electron optics. Here only spherical aberration shall be briefly dealt with, as it
is the most important in the resolving power of a lens.

3.6. Spherical Aberrations

The rays leaving the source at different inclinations u, do not
return to axis at the same point. Those with larger angle return a shorter
distance away. Thus, the image of a point is a disc in focal plane. This is called spherical aberration or apperture defect. It represents the most serious
limitation to resolving power of a lens.

A beam of rays within a cone of aperture γ and the radius of
lens zone h will produce a disc image, with the radius of the disc Δr_i
proportional to h^3.

$$\Delta r_i = Ch^3$$

where C is a constant of proportionality. (Fig. 3.3).

References

1. P.Grivet, Electron Optics, Pergamon, New York, 1965

2. W.Glaser, Die Grudlagen der Elektronenoptik, Springer, Wien, 1952

3. J.Picht, Einfuhrung in die Theorie der Elektronenoptik, Barth, Leipzig, 1939

Text to Figures. Ch.3

Fig. 3.1. Cardinal points of a lens: focal points F, F^1; nodal points N, N^1 and principal planes H, H^1.

Fig. 3.2. Derivation of angular magnification.

Fig. 3.2. The spherical aberration.

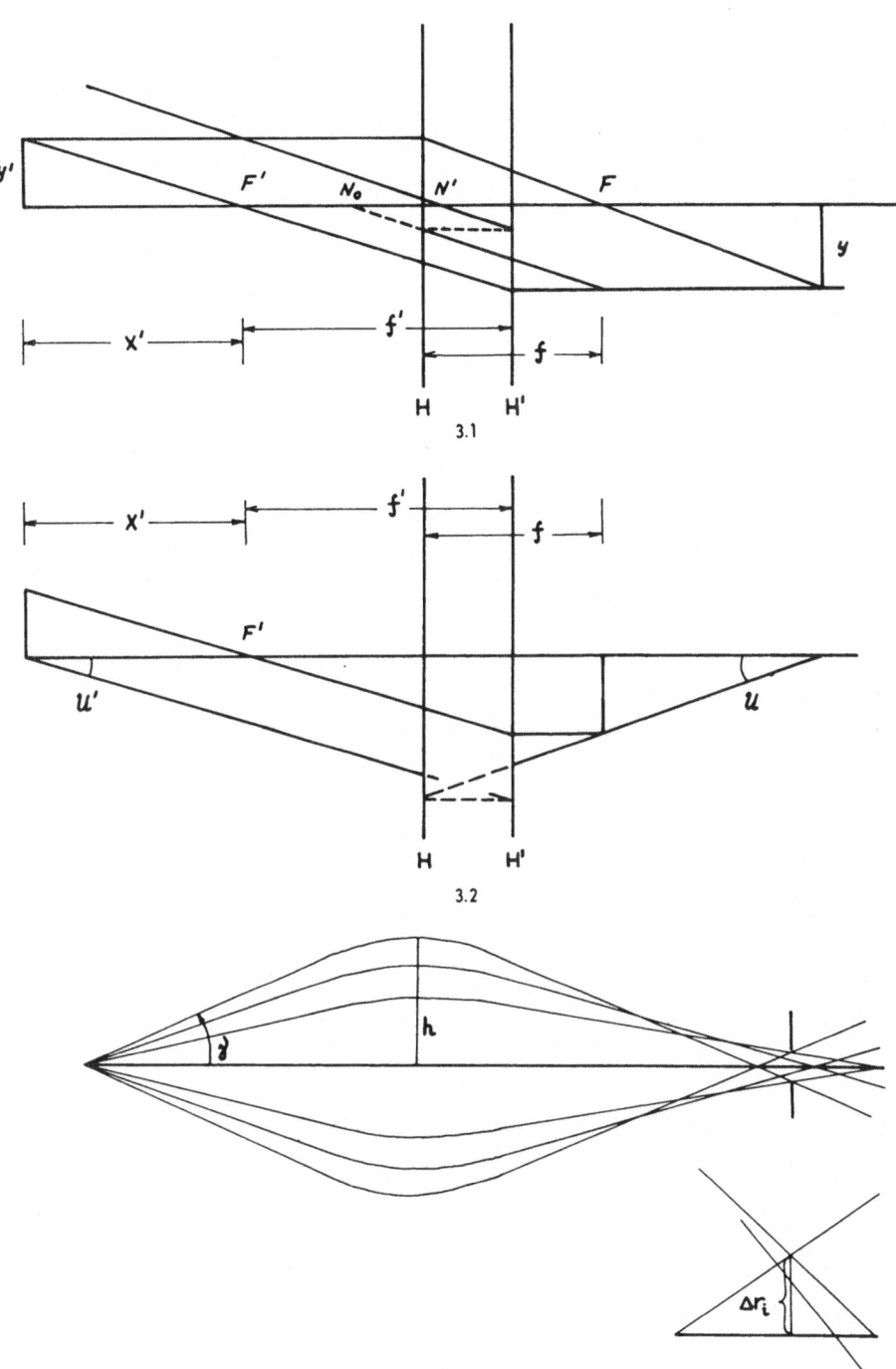

3.1

3.2

3.3

4. ELECTRON MOTION IN CYLINDRICAL FIELDS

In flat spectrometers, the magnetic field has a plane of symmetry, which contains the source and detector. The simplest apparatus of this type is semi-circular spectrometer, with homogenous field. More important, however, are the spectrometers with inhomogenous fields. Besides the reflection symmetry with respect to median plane, they usually possess other symmetries. Most of these fields have cylindrical symmetry with respect to an axis perpendicular to the median plane. The whole field is then defined by the field in the median plane, as it follows from Maxwell equations. The magnetic field in the median plane has only axial components and may be expressed in form of a series

$$B_z(r, O) = B_o \left[1 + \alpha_1 \left(\frac{r - r_o}{r_o} \right) + \alpha_2 \left(\frac{r - r_o}{r_o} \right)^2 + \ldots \right]$$

where r_o is the radius of the circle containing the centers of the source and the detector, which we shall call the optic circle and B_o is the field at the optic circle. The trajectories in the vicinity of the optic circle may be found by considering $(r - r_o)$ as a small quantity and solving the equations of motion by successive approximations.

In this chapter we shall consider the theory of electron motion in cylindrical fields, as developed by Svartholm (1) and Lee-Whiting (2).

Magnetic field components with axial symmetry – We shall find the components of an axially symmetric magnetic field in a cylindrical coordinate system. It is convenient to use the vector potential $\vec{A}$ defined by

$$\vec{B} = \text{curl } \vec{A}$$
$$\text{div } \vec{A} = O \tag{4.1}$$

The components of B in cylindrical coordinates are

$$B_r = \frac{1}{r} \left(\frac{\partial A_z}{\partial \theta} - \frac{\partial (rA_\theta)}{\partial z} \right)$$

$$B_z = \frac{1}{r} \left(\frac{\partial (rA_\theta)}{\partial r} - \frac{\partial A_r}{\partial \theta} \right) \tag{4.2}$$

$$B_\theta = \left(\frac{\partial A_r}{\partial z} - \frac{\partial A_z}{\partial r} \right)$$

In the case of an axially symmetric magnetic field, A is normal to both the axis of symmetry and to the radius vector from the axis to the point where A is considered. This follows from the fact that axially symmetric fields are produced by currents flowing circularly around the axis. Then only A_θ is different from zero (3) and we have

$$A_r = A_z = O \qquad \text{and} \qquad A_\theta = A$$

so that equations (4.2) reduce to

$$B_r = \frac{1}{r} \frac{\partial(rA)}{\partial z} = \frac{\partial A}{\partial z}$$

$$B_z = \frac{1}{r} \frac{\partial(rA)}{\partial r} \qquad\qquad\qquad\qquad (4.3)$$

$$B_\theta = O$$

Series expansion of vector potential – An electron emitted tangentially to the optic circle will move along optic circle if its momentum p_o is connected with B_o and r_o by the relation

$$B_o r_o = - p_o/e \qquad\qquad\qquad\qquad (4.4)$$

For the orbits near the optic circle, it is convenient to express the electron coordinates in the following form

$$\eta = (r - r_o)/r_o \qquad \tau = z/r_o \qquad\qquad (4.5)$$

Both η and τ are small quantities, usually smaller than 0,1.

The field can be expanded in terms of η and τ. Since the cylindrical field components (3) are simply derivatives of rA, the field expansion can be written in the form

$$rA = r_o A_o + B_o r_o^2 \sum_{m,n=o}^{\infty} C_{mn} \tau^m \eta^n \qquad (4.6)$$

where the following coefficients have values 0 or 1:

- Due to reflection symmetry in $z = o$ plane, all coefficients with m odd are zero.

- $c_{00} = 0$

- $c_{01} = 1$, in order to have proper field at optic circle defined by

(4).

In (6) the value of vector potential on the optic circle is denoted by A_0.

It is well known that magnetic field with cylindrical symmetry is determined by the field $B_z(r,o)$ in the median plane. When the field is defined by (6), coefficients c_{on} should be sufficient to find all the other coefficients c_{mn}. The relation between coefficients follows from Maxwell equation for cylindrical field in vacuo.(curl $\vec{B} = 0$)

$$\frac{\partial \dot{B}_r}{\partial \tau} - \frac{\partial B_z}{\partial \eta} = 0 \qquad (4.7)$$

The other equation (div B = o) being satisfied by the definition of vector potential.

Expressing B_r and B_z with help of (3) and (6), above equation gives the following recurrence formulae, when coefficients of $\tau^m \eta^n$ from the two sides are equaled

$$n \geq 0 \quad (m+2)(m+1)\left[c_{m+2,\,n+1} + c_{m+2,n}\right] = -(n+2)\left[(n+3)c_{m,\,m+3} + n\,c_{m,n+2}\right] \qquad (4.8)$$

$$(m+2)(m+1)c_{m+2,o} = -2c_{m\,2} + c_{m,1} \qquad (4.9)$$

The field in the median plane is usually defined by

$$B_z(r,O) = B_o \sum_{n=o}^{\infty} \alpha_n \eta^n \qquad (4.10)$$

where $\alpha_o = 1$. The connection between c_{on} and α_n can be obtained from (3) and (6), which together with (10) give

$$n\,c_{on} = \alpha_{n-1} + \alpha_{n-2} \qquad (4.11)$$

Equations of motion – Starting from Lorentz equation

$$\vec{F} = m\vec{a} = e\,(\vec{v} \times \vec{B})$$

which in cylindrical coordinates is expressed by

$$a_r = \ddot{r} - r\dot{\theta}^2 = -(e/m)r\dot{\theta}B_z \tag{4.12a}$$

$$a_z = \ddot{z} = -(e/m)r\dot{\theta}B_r \tag{4.12b}$$

$$a_\theta = \frac{1}{r}\frac{d}{dt}(r^2\dot{\theta}) = (e/m)\dot{r}B_z \tag{4.12c}$$

and introducing field components (3), equations of motion become

$$\ddot{r} - r\dot{\theta}^2 = -(e/m)\dot{\theta}\frac{\partial}{\partial r}(rA) \tag{4.13}$$

$$\ddot{z} = -(e/m)\dot{\theta}\frac{\partial}{\partial z}(rA) \tag{4.14}$$

$$\frac{\partial}{\partial t}(r^2\dot{\theta}) = (e/m)\frac{\partial}{\partial t}(rA) \tag{4.15}$$

Equation (15) can be integrated immediately to give

$$r\dot{\theta} = (e/m)(rA) + const \tag{4.16}$$

The constant of integration can be found from initial conditions, which we put, following Lee-Whiting's notation

$$\eta = h \qquad \tau = t \qquad \theta = 0$$

$$\frac{d\eta}{d\theta} = \eta' = H \qquad \frac{d\tau}{d\theta} = T \tag{4.17}$$

γ = angle between the initial momentum vector and the tangent to the optic circle at $\theta = 0$, given by

$$\cos \gamma = \left| 1 + (H^2 + T^2)/2(1 + h)^2 \right|^{-1/2} \tag{4.18}$$

$$p = p_0(1 + \varepsilon) = \text{absolute value of electron momentum} \tag{4.19}$$

Since $r\dot{\theta} = (p/m)\cos\gamma$, equation (16) can be written as

$$r(p/m)\cos\gamma = (e/m)(rA) + const \tag{4.20}$$

Using (5), (6), (17), (18), and (19), equation (20) becomes

$$(r_0 p_0/m)(1 + \eta)(1 + \varepsilon)\cos\gamma = (e/m)\left[r_0 A_0 + B_0 r_0^2 \sum_{m,n} c_{m,n} t^m h^n \right] + const$$

which with the help of (4) gives for value of const

$$\text{const} = -(e/m)B_o r_o^2 \left[(1+\eta)(1+\epsilon)\cos\gamma + \Sigma c_{mn} t^m h^n \right] - (e/m)r_o A_o$$

and equation (16) becomes

$$r^2 \dot\theta = (e/m)B_o r_o^2 \left[(1+\eta)(1+\epsilon)\cos\gamma + \Sigma c_{mn}(\tau^m \eta^n - t^m h^n) \right] \qquad (4.21)$$

Introducing F defined as

$$F = (1+\eta)(1+\epsilon)\cos\gamma + \Sigma c_{mn}(\tau^m \eta^n - t^m h^n) \qquad (4.22)$$

equation (21) becomes, using also (5)

$$\dot\theta = (e/m)B_o (1+\eta)^{-2} F \qquad (4.23)$$

In order to solve (13) and (14) it is convenient to change indepen-dent variable from t to θ. Denoting

$$\frac{dr}{d\theta} = r', \quad \frac{d^2 r}{d\theta^2} = r'', \quad \text{it is easy to find that}$$

$$\ddot r = \dot\theta^2 r'' + \dot\theta \frac{d\dot\theta}{d\theta} r' \qquad (4.24)$$

We need first to find $\dfrac{d\dot\theta}{d\theta}$, which with the help of (23) becomes

$$\frac{d\dot\theta}{d\theta} = \frac{d\dot\theta}{dr} r' + \frac{d\dot\theta}{dz} z'$$

$$= (e/m)B_o(1+\eta)^{-2}\left\{ \eta' \left[P - 2(1-\eta)^{-1} F \right] + \tau' Q \right\} (4.25)$$

where

$$P \equiv \frac{\partial F}{\partial \eta} \qquad \text{and} \quad Q \equiv \frac{\partial F}{\partial \tau} \qquad (4.26)$$

We will also need

$$\frac{1}{\dot\theta} \frac{\dot\theta}{\theta} = \frac{1}{F}\left\{ \eta' \left[P - 2(1+\eta)^{-1} F \right] + \tau' Q \right\} \qquad (4.27)$$

Using (24), equation (13) becomes

$$\dot{\theta}^2 r'' + \dot{\theta} \frac{d\dot{\theta}}{d\theta} r' - \dot{\theta}^2 r = -(e/m)\dot{\theta} \frac{\partial}{\partial r} (rA) \qquad (4.28)$$

Dividing by $\dot{\theta}^2$ and passing to new independent variables, (28) can be written as

$$\eta'' + \frac{1}{\dot{\theta}} \frac{d\dot{\theta}}{d\theta} \cdot \eta' - (1+\eta) = (e/m) \frac{1}{r_o^2} \frac{1}{\dot{\theta}} \frac{\partial}{\partial \eta} (rA) \qquad (4.29)$$

Introducing (27) and taking into account that

$$\frac{\partial}{\partial \eta} (rA) = B_o r_o^2 \frac{\partial F}{\partial \eta} \qquad (4.30)$$

equation (29) becomes

$$\eta'' + \eta' \left\{ \frac{P}{F} \eta' + \frac{Q}{F} \tau' - \frac{2\eta'}{1+\eta} \right\} = -\frac{P}{F} (1+\eta)^2 + 1 + \eta \qquad (4.31)$$

Noting that equation (14) has a similar structure to (13) except for term $r\dot{\theta}^2$ which contributes in (31) the term $(1+\eta)$, it is easy to see that (14) becomes

$$\tau'' + \tau' \left\{ \frac{P}{F} \eta' + \frac{Q}{F} \tau' - \frac{2}{1+\eta} \eta' \right\} = -\frac{Q}{F} (1+\eta)^2 \qquad (4.32)$$

Equations (31) and (32) differ from Lee-Whitings equations (9) and (10) by factor 2 on the right side, which is due to change of variable $\theta = \sqrt{2}\,\Psi$, which he made for convenience in handling $\pi\sqrt{2}$ spectrometer calculations.

First order solutions – First order solutions are obtained by assuming that h, t, H, T and ϵ are all small compared to unity and retaining only the terms of first order in equations (31) and (32). These equations are then reduced to

$$\eta'' + 2c_{o2}\, \eta = \epsilon \qquad (4.33)$$

$$\tau'' + 2c_{2o}\, \tau = 0 \qquad (4.34)$$

From (11) one finds that

$$2c_{o2} = 1 + \alpha_1 \qquad (4.35)$$

while (9) for $m = 0$ reduces to

$$2 c_{02} + 2 c_{20} = 1 \qquad (4.36)$$

which, together with (35), gives

$$2 c_{20} = - \alpha_1 \qquad (4.37)$$

Using (35) and (37) equations (33) and (34) become

$$\eta'' + (1 + \alpha_1) = 0 \qquad (4.38)$$

$$\tau'' - \alpha_1 \tau = 0 \qquad (4.39)$$

where we have also set $\varepsilon = 0$ because we will first consider the focusing of electrons with momentum p_0.

The sign and magnitude of α_1 determines whether the solutions are sinusoidal or exponential. Focalisation is obtained for sinousoidal solutions which requires that coefficients with η and τ should be positive. It follows then from (39) that

$$\alpha_1 < 0$$

and this requires that in (38) the coefficient satisfies

$$1 + \alpha_1 > 0$$

Both inequalities can be written

$$0 > \alpha_1 > - 1 \qquad (4.40)$$

Since coefficients with τ and η have to be positive it is convenient to introduce

$$\omega_r^2 = 1 + \alpha_1 \qquad (4.41)$$

$$\omega_z^2 = - \alpha_1 \qquad (4.42)$$

Equations (33) and (34) then become

$$\eta'' + \omega_r^2 \eta = 0 \qquad (4.43)$$

$$\tau'' + \omega_z^2 \tau = 0 \qquad (4.44)$$

The general solutions of these equations are

$$\eta = C_1 \cos \omega_r \theta + C_2 \sin \omega_r \theta \qquad (4.45)$$

$$\tau = C_3 \cos \omega_z \theta + C_4 \sin \omega_z \theta \qquad (4.46)$$

The constants of integration are determined from initial conditions. If an electron starts from the point on the optic axis defined by

$$h = o, \quad t = o, \quad \theta = o$$

the constants C_1 and C_3 are zero.

The other two constants are obtained by taking derivations with respect to θ, which should be equal to H and T, defining radial and axial apertures. We have then

$$\eta' = C_2 \, \omega_r = H \qquad \tau' = C_4 \, \omega_z = T$$

or

$$C_2 = H/\omega_r \qquad C_4 = T/\omega_z$$

so that first order solutions are

$$\eta = (H/\omega_r) \sin (\omega_r \theta) \tag{4.47}$$

$$\tau = (T/\omega_z) \sin (\omega_z \theta) \tag{4.48}$$

It is convenient for higher order calculations to introduce

$$H_o = H/\omega_r \qquad T_o = T/\omega_z$$

so that (47) and (48) are written as

$$\eta = H_o \sin (\omega_r \theta) \tag{4.49}$$

$$\tau = T_o \sin (\omega_z \theta) \tag{4.50}$$

It is easily seen that H_o and T_o are the values of η and τ for half of the focusing angle, where the departure of the beam from optic axis is maximum. Fig.4.1.

The general form of first order solutions for electrons starting from point (h, t) is

$$\eta = H_o \sin (\omega_r \theta) + h \cos (\omega_r \theta) \tag{4.51}$$

$$\tau = T_o \sin (\omega_z \theta) + t \cos (\omega_z \theta) \tag{4.52}$$

<u>Focusing angles</u> - A particle leaving the optic circle at $\theta = 0$ shall have periodically the values $\eta = o$ and $\tau = o$ for the angles $\phi_r^{(m)}$ and

$\emptyset_2^{(n)}$ given by

$$\emptyset_r^{(m)} = m\,\pi/\omega_r = m\,\pi\,(1-\alpha_1)^{-1/2} \tag{4.53}$$

$$\emptyset_z^{(n)} = n\,\pi/\omega_z = n\,\pi\,(-\alpha_1)^{-1/2} \tag{4.54}$$

In the general case the radial focusing angle $\emptyset_r^{(m)}$ do not coincide with the axial focusing angle $\emptyset_z^{(n)}$. They are functions of the magnetic field shape, characterised in the first order approximation by α_1.

An analysis given further below will show that the dispersion is maximum for $m = 1$, being zero for even values of m. For that reason we shall take $m = 1$, which gives

$$\emptyset_r = (1-\alpha)^{-1/2} \tag{4.55}$$

First radial focusing angle $\emptyset_r$ is connected with the first axial focusing angle $\emptyset_z = \pi/\sqrt{-\alpha_1}$ by the relation

$$\frac{1}{\emptyset_r^2} + \frac{1}{\emptyset_z^2} = \frac{1}{\pi^2} \tag{4.56}$$

which is obtained by squaring and adding the inverse values of $\emptyset_r$ and $\emptyset_z$. The increase of one focusing angle is accompanied by a decrease of the other. Since the dispersion increases with square of the radial focusing angle, as we shall see further below (70), larger radial angles are of interest. It is possible then that n can have values larger than 1. Let us consider some specific cases.

$\underline{\alpha_1 = 0}$ - For a homogenous field $\alpha_1 = 0$, which gives

$$\emptyset_r = \pi \qquad \emptyset_z = \infty \tag{4.57}$$

This corresponds to the case of semi-circular spectrometers with radial focusing and no axial focusing.

Double focusing (DF) - A two directional focus will be obtained at an angle $\emptyset_0$ given by

$$\emptyset_0 = (1-\alpha_1)^{-1/2} = n\,\pi\,(-\alpha_1)^{-1/2} \tag{4.58}$$

from which follows

$$\alpha_1 = - n^2/(1 + n^2) \tag{4.59}$$

$$\phi_0 = \pi \sqrt{1 + n^2} \tag{4.60}$$

For few first integers, relations (59) and (60) give the following values for DF spectrometers:

TABLE 4.I.

n	α_1	ϕ_0
1	- 0.5	$\pi\sqrt{2} = 255^o$
2	- 0.8	$\pi\sqrt{5} = 402^o$
3	- 0.9	$\pi\sqrt{10} = 569^o$

The value of n = 1 corresponds to the well known case of Sieg-bahn - Swartholm double focusing spectrometer. For n > 1 focusing angles become larger than 360°. The beam would first have to by-pass the detector, then after passing over the source to enter the detector. Some of the beam has to be lost. We shall return to this question in more detail later.

Extended focus (EF) spectrometers - Daniel pointed out (4) that it is possible to improve the resolution by giving up the first order axial focusing and accepting an extended focus placed at $\phi_z (2N - 1)/2$. The focusing angles ϕ_0 are then given by the relation

$$\phi_0 = (\frac{2N - 1}{2}) \frac{\pi}{\sqrt{-\alpha}} = \frac{\pi}{\sqrt{1+\alpha_1}} \qquad N - \frac{1}{2} = n \tag{4.61}$$

from which follows

$$\alpha_1 = - (2N - 1)^2/ \left[(2N - 1)^2 + 4 \right]$$

$$\phi_0 = (\pi/2) \sqrt{(2N - 1)^2 + 4} \tag{4.62}$$

For the first few integers the following values of α_1 and ϕ_0 are obtained for EF spectrometers.

TABLE 4.II

N	n	α_1	ϕ_o
1	1/2	- 0.2	$(\pi/2)\sqrt{5} = 202^o$
2	3/2	- 0.69	$(\pi/2)\sqrt{13} = 325^o$
3	5/2	- 0.86	$(\pi/2)\sqrt{29} = 485^o$

For $N = 2$ the focusing angle is still smaller than 360°, while for higher values of N, ϕ_o increases beyond 2π .

Dispersion - In order to find the dispersion it is necessary to consider the motion of electrons with momentum $p = p_o (1 + \varepsilon)$, described in the first order by equations (33) and (34). The radial equation can be written as

$$\eta'' + \omega_r^2 \eta = \varepsilon \tag{4.64}$$

The general solution of (64) is

$$\eta = C_1 \cos(\omega_r \theta) + C_2 \sin(\omega_r \theta) + \varepsilon / \omega_r^2 \tag{4.65}$$

When C_1 and C_2 are found from the same initial conditions as were used for solution of (43), one obtains

$$\eta = H_o \sin(\omega_r \theta) + (\varepsilon / \omega_r^2) \left[1 - \cos(\omega_r \theta)\right] \tag{4.66}$$

In the θ plane of radial focusing, where

$$\theta = \phi_r = \pi / \omega_r$$

(66) has the value

$$\eta_o = \frac{2\varepsilon}{1 + \alpha_1} \tag{4.67}$$

Since

$$\eta_o = \frac{r - r_o}{r_o} = \frac{dr}{r_o} \tag{4.68}$$

$$\varepsilon = \frac{p - p_o}{p_o} = \frac{dp}{p} \tag{4.69}$$

dispersion D_r, defined as

$$D_r = \frac{dr}{r} \frac{P_o}{dp}$$

becomes

$$D_r = \frac{2}{1 - \alpha_1} = \frac{2}{\pi^2} \emptyset_r^2 = (2 (1 + n^2) \text{ for DF}) \tag{4.70}$$

It is seen that dispersion increases with the square of radial focusing angle, as illustrated in the table given below

TABLE 4.III.

α_1	$\emptyset_r$	D_r	Type of focusing
0	180°	2	Semi - circular
- 0.2	202°	2.5	EF
- 0.5	255°	4	Siegbahn – Swartholm DF
- 0.69	325°	6.5	EF
- 0.8	402°	10	DF
- 0.86	485°	14	EF
- 0.9	569°	20	DF

<u>Second order solutions</u> – Second order solutions are

$$\eta(\theta_o) = - h - (1 - \alpha_1)^{-1}(1 + \frac{7}{3} \alpha_1 + \frac{4}{3} \alpha_2)H_o^2 + (1 - \alpha_1)^{-1}(\alpha_1 + \mu\alpha_2)T_o^2$$

$$- \frac{3}{2}(1 + \alpha_1)^{-1}(\alpha_1 + \alpha_2)h^2 + (1 + \alpha_1)^{-1}(\alpha_1 + 2\alpha_2 - \mu\alpha_2)t^2 + \nu\alpha_2 T_o t \tag{4.71}$$

$$\tau(\theta_o) = 2(1 + 5\alpha_1)^{-1} \left\{ -\alpha_1(1 + \alpha_1)^{-1} \right\}^{1/2} (1 + 5\alpha_1 + 4\alpha_2)H_o T_o \cos X$$

$$+ 4 \alpha_2(1 + 5\alpha_1)^{-1} ht \cos X - h T_o \sin X \tag{4.72}$$

$$+ 2 (\alpha_1 + \alpha_2) \left\{ -\alpha_1(\alpha_1 + \alpha_2) \right\}^{-1/2} H_o t \sin X + t \cos X + T_o \sin X$$

where
$$\mu = (1 + 5\,\alpha_1)^{-1} \left\{ (1 + \alpha_1)\sin^2\chi + 4\,\alpha_1 \right\} \tag{4.73}$$

$$\nu = (1 + 5\,\alpha_1)^{-1} \sin 2\,\chi \tag{4.74}$$

$$\chi = \left\{ \alpha_1 (1 + \alpha_1)^{-1} \right\}^{1/2} \pi \tag{4.75}$$

The relation between the resolution and transmission depends primarily on terms in H_o and T_o in the expression for radial aberrations (71). The coefficients with H_o^2 and T_o^2 cannot simultaneously vanish. If the values of α_1 and α_2 are chosen to make the coefficient of T_o^2 in (71) equal to zero, larger axial opening can be used. This is often called high aperture spectrometer. On the other hand for wide-aperture spectrometer the coefficient of H_o^2 has to be equated to zero. It can be noted also that the coefficient of $H_o T_o$ in (72) vanishes for the same values of α_1 and α_2 as the coefficient of T_o^2.

Solid angle – The entrance baffle defines the solid angle of the accepted beam. In principle, entrance baffle should be shaped to minimize the aberrations. In the theoretical treatment of different types, especially when it is limited to second order, it is convenient to deal with rectangular entrance baffles, supposed to be placed at a radial angle corresponding to half of the radial focusing angle. The dimensions of the baffle are then

$$2\,r_o H_{om} \times 2\,r_o T_{om} \tag{4.76}$$

where we have denoted by H_{om} and T_{om} the maximum values of H_o and T_o respectively.

One should mention that in actual spectrometers the entrance baffles are not necessarily placed at half of the focusing angle, but for the purpose of analysis it is simpler to deal with equivalent "half-way" baffles.

The radial opening angle ζ_r and axial opening angle ζ_z are connected with H_{om} and T_{om} by relations

$$\tan \zeta_r = \left(\frac{d\eta}{d\theta}\right)_m = H_m = \omega_r H_{om} \tag{4.77}$$

$$\tan \zeta_2 = (\frac{d\tau}{d\theta})_m = T_m = \omega_z T_{om} \tag{4.78}$$

Since angles are small ones, (77) (78) can be written as

$$\zeta_r = \omega_r H_{om} \tag{4.79}$$

$$\zeta_z = \omega_z T_{om} \tag{4.80}$$

The solid angle of the accepted beam Ω then becomes

$$\Omega = \frac{2\zeta_r \cdot 2\zeta_z}{4\pi} \cdot 100\%$$

$$= \frac{\omega_r \omega_z H_{om} T_{om}}{\pi} 100\%$$

$$= \frac{\sqrt{(1 - \alpha_1)(-\alpha_1)}}{\pi} H_{om} T_{om} \%$$

$$= \frac{n}{\pi(1 + n^2)} \cdot H_{om} T_{om} \% \tag{4.81}$$

In the case of $\pi\sqrt{2}$ spectrometer, the solid angle is

$$\Omega = \frac{1}{2\pi} H_{om} T_{om} 100\% \tag{4.82}$$

When n increases, keeping H_{om} and T_{om} the same, solid angle is reduced. This reduction of the solid angle is more than compensated by increase of dispersion allowing the use of larger sources. Since the dispersion increases as $(1 + n^2)$, one can expect that the luminosity will increase at least as n. Correct treatment requires the consideration of second and higher order terms. Lee-Whiting calculated luminosity for the case when H_{om} vanishes up to the third order. Then for high resolutions, luminosities increase roughly as cube of n.

Wide aperture optimum field — Optimum values of the coefficients α have been calculated for the wide aperture case by Saulit (5) and by Huster et al. (6). The values of the coefficients are expressed in terms of a parameter C which is connected with focusing angle ϕ_f by the relation

$$C = (\pi / \phi_f)^2$$

The coefficients are

$$\alpha_1 = C - 1$$
$$\alpha_2 = -7/2C + 1$$
$$\alpha_3 = 19/8\,C - 1/2\,C^2 - 1$$
$$\alpha_4 = -187/64\,C + 87/64\,C^2 + 1$$
$$\alpha_5 = 437/128\,C - 1587/640\,C^2 + 9/40\,C^3 - 1$$

In the case of the $\pi\sqrt{2}$ spectrometer, the optimum field for wide aperture is obtained by putting $C = \dfrac{1}{2}$

$$B(\eta, 0) = B_0(1 - 1/2\,\eta + 1/8\,\eta^2 + 1/16\,\eta^3 - 31/256\,\eta^4 + 59/512\,\eta^5) \qquad (4.83)$$

This field was calculated by Pavinski (7) and soviet physicists call it by his name. It was also found by Verster (8).

For $C = 1$, one obtains a field previously found by Beiduk and Konopinski (9)

$$B(\eta, 0) = B_0(1 - 3/4\,\eta^2 + 7/8\,\eta^3 - 9/16\,\eta^4 + 51/320\,\eta^5) \qquad (4.84)$$

References Ch.4

1. N.Svartholm, Arkiv Mat. Astron. Fysik 33 A, No 24, (1946) and Arkiv Fysik 2, No 20 (1949)

2. G.E. Lee-Whiting and E.A.Taylor, Can. J. Phys. 35, 1 (1957)

3. See for example: Ernst Weber: "Electromagnetic Fields", Chapman Hall, London, 1950

4. H.Daniel, RSI 31, 249 (1960)

5. V.R.Saulit, Izv.Akad.Nauk SSSR, Ser.Fiz. 18, 227 (1954)

6. Huster, Lehr and Walcher, Z.Naturforsch. 10a, 83 (1955)

7. P.P.Pavinskii, Izv. Akad. Nauk SSSR, Ser. Fiz. 18, 175 (1954)

8. N.F.Verster, Physica 16, 815 (1950)

9. F.Beiduk and E.Konopinski, RSI 19, 594 (1948)

Text to Figures. Ch.4

Fig. 4.1. The coordinate system and notation for mathematical treatement
of cylindrical fields.

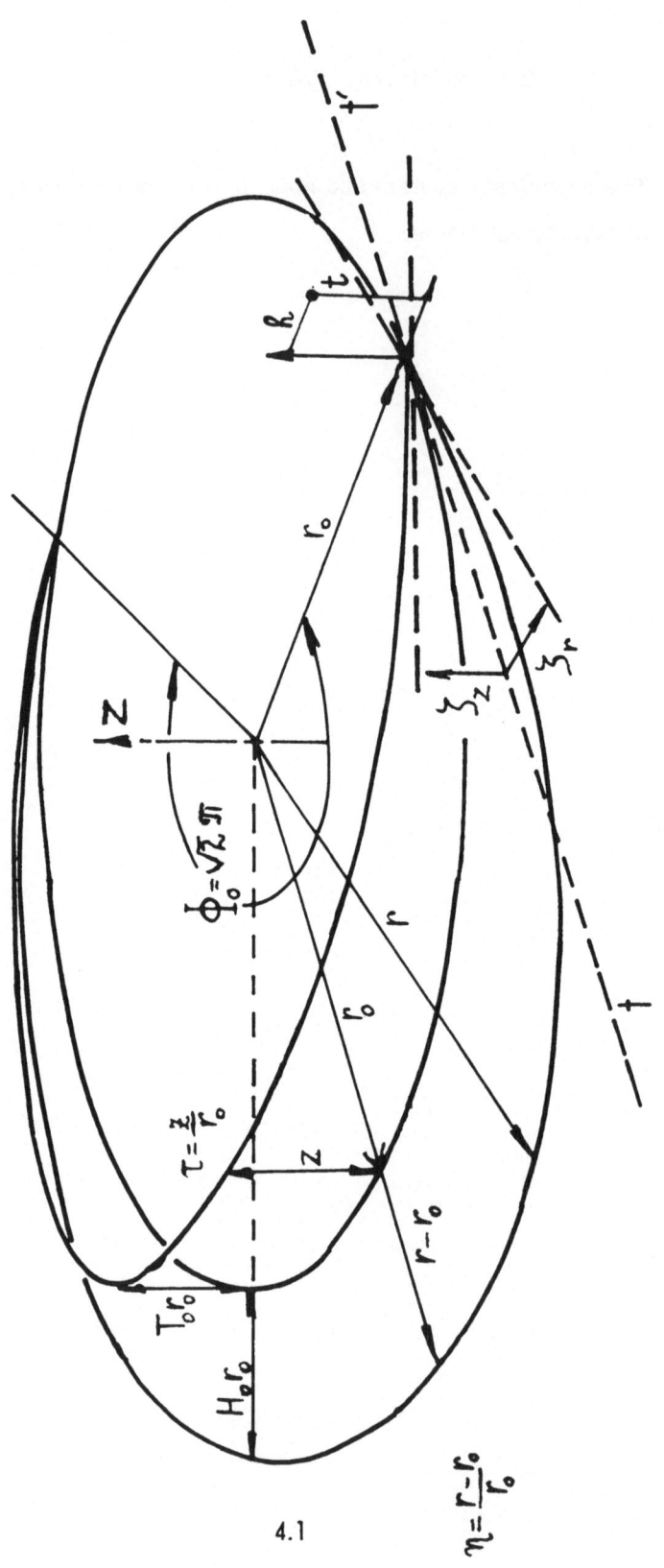

4.1

5. SEMI–CIRCULAR SPECTROMETERS

Beam geometries - There are three principal beam geometries, illustrated in Fig.5.1a, b, and c.

Constant radius geometry is shown in Fig.5.1a. Entrance and exit slit are fixed. Spectrum is obtained by changing the magnetic field. Obviously, permanent magnet is unsuitable for this type of geometry.

Constant field, I–geometry (Fig.5.1b) has the source aligned with an extended detector. In this case it is relatively easy to make the energy measurements of electron lines. The energy of a line may be defined by the central ray C, describing exactly the semi–circle and forming the high energy end of the image. The other two principal rays, L and R, passing just off the slit edges, fall nearer to the side of the source. This is valid, however, only for an intermediate range of radii.

Constant field, Γ - geometry has the source–slit line forming with focal line a Greek letter Γ. In this case central ray always forms the high energy edge, while the other two principal rays always fall into the same point, forming the low energy edge of the image of the point source. One can also see how the width of the image increases with the radius This was up to now the most widely used geometry of semi–circular spectrometer.

With constant field spectrometers each of the three types of magnets can be used, but permanent magnet is the most convenient.

Line shape - Many authors have studied the shape of a monochromatic line produced by an extended source (1-15). Earlier authors studied especially the cyllindrical sources because the most important sources of the early period were radioactive deposits of radon and thoron, which were collected on wires. Source–supports of artifically radioactive elements are usually very thin flat metalised foils. The foil is preferred to wire because the low energy tail produced by scattering is larger when wire is used as support. The later authors (3-15) studied the line shape of a flat rectangular source.

We shall give here a two–dimensional graphic treatment of line shape formation, due to Cork and collaborators (7). The two–dimensional

treatment has the advantage of being simple, and yet preserving essential features of more exact line shape formation.

Fig. 5.2 shows the construction of an area A formed by intersections of the four circles, all with the same radius R, centered at the ends of a
linear source s', s" and the edges of entrance diaphragme d', d". Area A
contains the centers of all the particle trajectory circles of radius R which
can originate from the source and pass through the entrance slit.

The relative intensity of electrons falling between two adjacent
points x', x" on the focal line is proportional to the area enclosed in A by
two arcs inscribed from x' and x". Authors also give an exact analytical
treatment.

We reproduce also three graphs showing the dependence of line
shape with radius, source width and entrance slit width. One can see from
Fig. 5.3 that when the radius is increasing the line becomes wider and the
height decreases. Figs. 5.3 – 5.5 show how important it is to match the relevant parameters, since the increase of source width or entrance slit beyond
certain optimum values may seriously reduce the resolution, while the intensity gain remains very small.

<u>Performance parameters</u> – We shall start with instrumental
contributions to the width of the image. The first, Δx_ρ is due to the radial
opening of the beam 2ζ . Fig. 5.6 shows three principal rays defining Δx_ρ .
It is seen that Δx_ρ is determined by the points where the rays C and L fall
on the focus. From properties of triangles inscribed in the circles one obtains

$$\Delta x_\rho = 2\rho - 2\rho \cos\zeta$$

$$= \rho\zeta^2 \tag{5.1}$$

In the above derivation $\cos\zeta$ was developed in the series and only
the first two terms, taken because ζ is usually very small (of the order of 0.1
radian).

The length of the source and the detector slit contribute to the image width an amount x_ψ , which will depend on the length of the source h, the
length of detector slit and the radius of the trajectory ρ . The length of the

detector slit may differ much according to the type of spectrometer. If a photographic plate is used the microphotometer slit will usually have a small length, while a Geiger counter would have a slit length about the same as the length of the source. We will suppose that the length of the detector slit is the same as the length of the source h.

The largest lateral opening ψ may be defined as the angle between two rays leaving the one end of the source and falling at two extreme ends of the detector slit.

Taking h to be a half of an helix pitch, and since ψ is small, one finds that

$$\psi = h/\pi\rho \tag{5.2}$$

Opening angle ψ decreases for larger radii.

An electron leaving from the point $(x = o, z = -h/2)$ in the plane z - x, with an angle ψ with respect to y-axis, will make half of an helix and hit z-x plane at a point having the coordinates equal to $(z = +h/2, x = 2 \cos\psi)$. This produces an image width equal to

$$\Delta x_\psi = 2\rho - 2\rho \cos \psi$$
$$= \rho\psi^2 \tag{5.3}$$

since ψ is very small.

Another contribution to the instrumental width comes from the width of the exit slit w, so that the total width of the image Δx is equal to

$$\Delta x = s + w + \rho\zeta^2 + \rho\psi^2 \tag{5.4}$$

Starting foom the total width of the image, one can obtain the expression for relative half-width by

1. Noting that in semi-circular spectrometers, one has the relation

$$\Delta x = 2 \Delta\rho$$

2. Assuming that the width at half-height is equal to one half of the width at the base.

Instrumental half-width, or resolution becomes then

$$R = \frac{\Delta(B\rho)}{B\rho} = \frac{\Delta x}{4} \cdot \frac{1}{\rho} = \frac{s+w}{4\rho} + \frac{1}{4}\,(\zeta^2 + \psi^2) \qquad (5.5)$$

Equations 5.4 and 5.5 illustrate the basic limitation of semi-circular spectrometers. They contain squares of radial and axial opening angles, which seriously limit the transmission for a given resolution.

The solid angle Ω is defined by aperture angles $2\,\zeta$ and ψ , so that it is given by

$$\Omega = 2\zeta\psi/4\pi = \zeta\psi/2 \qquad (5.6)$$

The geometric luminosity L_g becomes

$$L_g = sh\,\zeta\psi/2\pi \qquad (5.7)$$

The dispersion D was given earlier (Table 4.III), as

$$D = 2\rho$$

The expressions are valid for all three beam geometries. It is instructive to consider how they vary with the radius.

In the expression for total base width (5.4) the term $\rho\zeta^2$ increases, while $\rho\psi^2$ decreases, since ψ is inversely proportional to ρ . Resolution improves at higher ρ due to the factor $1/\rho$ and also ψ^2 is diminishing. Transmission is reduced at higher ρ because of the factor ψ in the formula for Ω (5.6).

Optimisation - In each experiment it is necessary to match various parameters, such as slits and source size, to obtain the expected total intensities and obtain resolutions in minimum time. Quite often the choice of parameters represents a compromise between the desired resolution and reasonable counting rate. Several authors (14-16) have calculated the optimum parameters for different cases, using various approaches. The conditions of the experiment may vary so much from one case to another that it is hardly possible for unique optimum relations between parameters to exist. We will only quote the results of calculations made by Geoffrion (16), for the case of "high source" giving optimum luminosity, for a given resolution. He obtains the following relations where we have for the sake of simplicity rounded off the numerical values (putting $7/4 \approx 2$):

$$\text{source widths} = \text{exit slit width}\quad w = \rho R$$

where R is defined by (5.5)

height of source and exit slit $h = \rho \pi \sqrt{R}$

radial half-angle $\zeta = \sqrt{R/2}$

axial opening $\psi = \sqrt{R}$

geometric luminosity $L_g = 0.35 \cdot \rho^2 R^{5/2}$

solid angle $\Omega = R/10$

$$h/s = \pi / \sqrt{R}$$

Let us illustrate the above relations for the case of high and low resolutions. $(R = 10^{-3}$ and 10^{-4}, $\rho = 50$ cm$)$

	$R = 10^{-3}, \rho = 50$ cm	$R = 10^{-4}, \rho = 50$ cm
s, w	0.5 mm	0.05 mm
h	50 mm	16 mm
ζ	2.10^{-2}	7.10^{-3}
ψ	3.10^{-2}	10^{-2}
Ω	10^{-4}	10^{-5}
L_g	3.10^{-7} cm^2	9.10^{-8} cm^2

The table shows that at very high resolutions an impossibly thin source is required and only at half-widths larger than 10^{-3} the width of the source approaches practical dimensions. Solid angles, also, are extremely small. Geoffrion has constructed a semi-circular spectrometer (12) based on the above optimisation, with a working resolution of $2.5 \cdot 10^{-3}$ and a radius of 30 cm.

Wide beam focusing - Balodis, Bondarenko, Prokoffiev and Servons (17-18) have proposed the use of magnetic screening tubes partially introduced into the pole-pieces gap, to allow the focusing of wider beams. The idea is to shape the inner end of the screening tube to correct for aberrations (fig. 5. 7). The calculations were made neglecting the end effects and assuming that the field is zero inside the tube and homogenous outside of the tube end boundary. Measurements have shown, of course, that this assumption is not correct. Measurements gave in some cases line-widths 80% larger than predicted by simple theory, but still several times smaller than what simple focusing would give.

A semi-circular spectrograph with correcting screening tube was developed for measurements of internal conversion emitted in (n, γ) reactions (14). Radiator-source had a diameter of 6 cm and was placed at a distance of 3 m from the spectrometer. Parallel beam of electron was focused on a photographic plate 70 cm long (Fig. 5. 8). The line width was 0.3-0.4%.

Spectrometers. - Most of the semi-circular spectrometers built so far, are of the constant field-variable radius type, with permanent magnet. They are simple to construct, and simple to operate, especially when photographic plate is used as detector. It is more convenient to use them for the identification of spectra and the energy measurements, while precise determination of intensities is very difficult, due to the non-linearity of the photographic blackening-intensity curve. Precise intensities can be obtained with more laborious method of track counting in the nuclear emulsions (19-24).

The first permanent m agnet spectrometer was built by Ellis, Cockroft and Kershaw (25), and used by Ellis for a series of measurements of radon, thoron and actionon. The Cambridge group payed particular attention to factors contributing to the precision of intensity and energy determinations. Their work marks the beginning of high precision beta-ray spectroscopy.

Further contribution to the precision was made by Slätis, who built a precisely mashined spectrometer, enclosed in a thermostated chamber (26), then added another two with fixed, complementary fields (27), and in a number of papers studied the photographic measurements of line intensities (28-31).

Sets of 3-4 spectrometers were built in several laboratories, especially those investigating weak sources with complex spectra, sometimes containing several isotopes (32-34). Since the fields are chosen so that the set covers continuously the spectrum up to several MeV, the field intensity of each instrument has to be set once for all. The construction can be then simplified. One of the simplest sets was recently designed by R. J.Wallen (35).

Mladjenović designed a relatively large instrument with pole-pieces having an area of $100x60$ cm^2, and the gap between them 6 cm wide, offering enough space for various interchangable detector arrangements (36). Either a $74x35$ cm^2 photographic plate could be used, or particle detectors moved from outside for scanning of the spectrum. Several large spectrometers were built (37-38) the largest, having maximum orbit radius of 75 cm, was recently constructed by Illes and Berenyi (39).

The first to buid large, alpha and beta spectrograph, whose work inspired others, was S.Rosenblum (40).

A list of semi-circular spectrometers, not mentioned above, is added for completeness in references (41-50).

References Ch.5

1. W.Wooster, Proc.Roy.Soc.A 114, 725 (1927)

2. K.T.Li, Proc. Combr. Phil. Soc. 33, 164 (1939)

3. R.Arnoult, Ann. de Phys. Ser. 11, t.12, 241 (1939)

4. R.J.Walen, J.Rech. C.N.R.S. 6, 156 (1955)

5. E.Cotton, Ann. Phys. 6, 481 (1951)

6. G.E.Owen, RSI 20, 916 (1949)

7. C.M.Fowler, R.G.Schreffler and J.M.Cork, RSI 20, 966 (1949)

8. C.G.Campbell and J.Kyles, Proc. Phys.Soc. B 66, 911 (1953)

9. C.M.Fowler, P.Domotor, J.Appl. Phys. 23, 415 (1952)

10. H.M.Kruse, G.P.Mellor and C.M.Fowler, J.Appl. Phys. 24, 1037 (1953)

11. E.Keberle, J.Res. C.N.R.S. No.49 (1959)

12. R.Stepić and M.Mladjenović, Bulletin "B.Kidrič" 10, No 220 (1960)

13. F.Schussler, Theses, Univ. Grenoble, 1965

14. J.L.Lawson and A.W.Tyler, RSI 11, 6 (1940)

15. K.Siegbahn, Ark. Mat. Fys. Astr. 30A, No 20 (1944)

16. C.Geoffrion, RSI 20, 638 (1949)

17. M.Balodis, V.Bondarenko, P.Prokoffiev, G.Sermons Izv. Akad. Nauk Latviiskoi SSR, No 11, 184 (1962)

18. M.Balodis, V.Bondarenko, P.Prokoffiev, Izv. Akad. Nauk SSSR, Ser.Fiz. 28, 263 (1964)

19. H.Slätis, NIM 2, 332 (1958)

20. K.D.Sevier, NIM 22, 345 (1963)

21. J.A.Antonova, Zhurn, Eksp. Teor. Fiz. 30, 571 (1956)

22. P.Z.Kleinheinz, Z.Naturforsh. 11, 252 (1956)

23. R.Carlson, A.Fahlman, R.Hallin and K.Siegbahn, Ark. Fys. 32, 99 (1966)

24. H.Slätis, Physica Scripta 7, 298 (1973)

25. J.D.Cockroft, C.D.Ellis and H.Kershaw, Proc. Roy. Soc. A 132, 442 (1931)

26. H.Slätis, Ark. f. Fys. 6, 415 (1953)

27. H.Slätis, NIM 2, 332 (1958)

28. H.Slätis, Ark.f. Fys. 8, 441 (1954)

29. M. Mladjenović and H. Slätis, Ark. f. Fys. 8, 65 (1954)

30. H. Slätis, Ark. f. Fys. 22, 517 (1962)

31. H. Slatis, NIM16, 51 (1962)

32. N. G. Smith and J. M. Hollander, Phys. Rev. 101, 746 (1965)

33. J. W. Mihelich, Phys. Rev. 87, 646 (1952)

34. A. Abdurazakov, F. Abdurazakova, K. Gromov, B. S. Dzhelepov, G. Ye.
    Umarov, Izv. Akad. Nauk, Uzbek. SSR 3, 53 (1961)

35. R. J. Walen, private communication

36. M. Mladjenović, Bull. Inst. "B. Kidrič" 6, 53 (1956)

37. J. Kormicki, H. Niewodniczanski, Z. Stahura, Nucleonika 11, 755 (1966)

38. N. Osis and A. Prokafierv, Izv. Akad. Nauk Latv. SSSR 9, 85 (1960)

39. F. Illes and D. Berenyi, Acta Phys. Hung. 18, 265 (1965)

40. S. Rosenblum, J. Santana, M. Valadares, J. Phys. Rad. 17, 112 (1956)

41. J. Surugue, Ann. de Phys. 8, 484 (1937)

42. A. F. A. Harper and N. F. Roberts, Proc. Roy. Soc. A 178, 170 (1941)

43. E. H. Plesset, G. P. Harnwell and F. G. P. Seidl, RSI 13, 351 (1942)

44. J. M. Cork, Phys. Rev. 72, 581 (1947)

45. R. D. Hill, Phys. Rev. 74, 78 (1948)

46. R. L. Caldwell, Phys. Rev. 78, 407 (1950)

47. G. E. Owen and C. S. Cook, RSI 20, 768 (1949)

48. N. Marty-Wallman, Ann. Phys. 6, 662 (1951)

49. R. Katz and M. R. Lee, RSI 25, 58 (1954)

50. E. Karlsonn and K. Siegbahn, NIM 7, 113 (1960)

Text to Figures. Ch. 5

Fig. 5.1.   The geometries of semi-circular spectrometers, (a) constant
radius, (b) constant field I type, (c) constant field Γ type.

Fig. 5.2.   The two-dimensional graphic treatment of line shape formation.
(Cork et.al.).

Fig. 5.3.   The increase of the line width for larger radii of electron
paths.

Fig. 5.4.   The dependance of the image width on the source width.

Fig. 5.5.   The dependence of the image width on the width of the entrance
slit.

Fig. 5.6.   The contribution of the radial opening to the image width.

Fig. 5.7.   Cylindrical magnetic shield with the inner profile curved to
improve the focusing of a parallel beam.

Fig. 5.8.   The semi-circular spectrograph with correcting screening tube for
measurements of internal conversion electrons emitted in (n, gamma)
reactions.

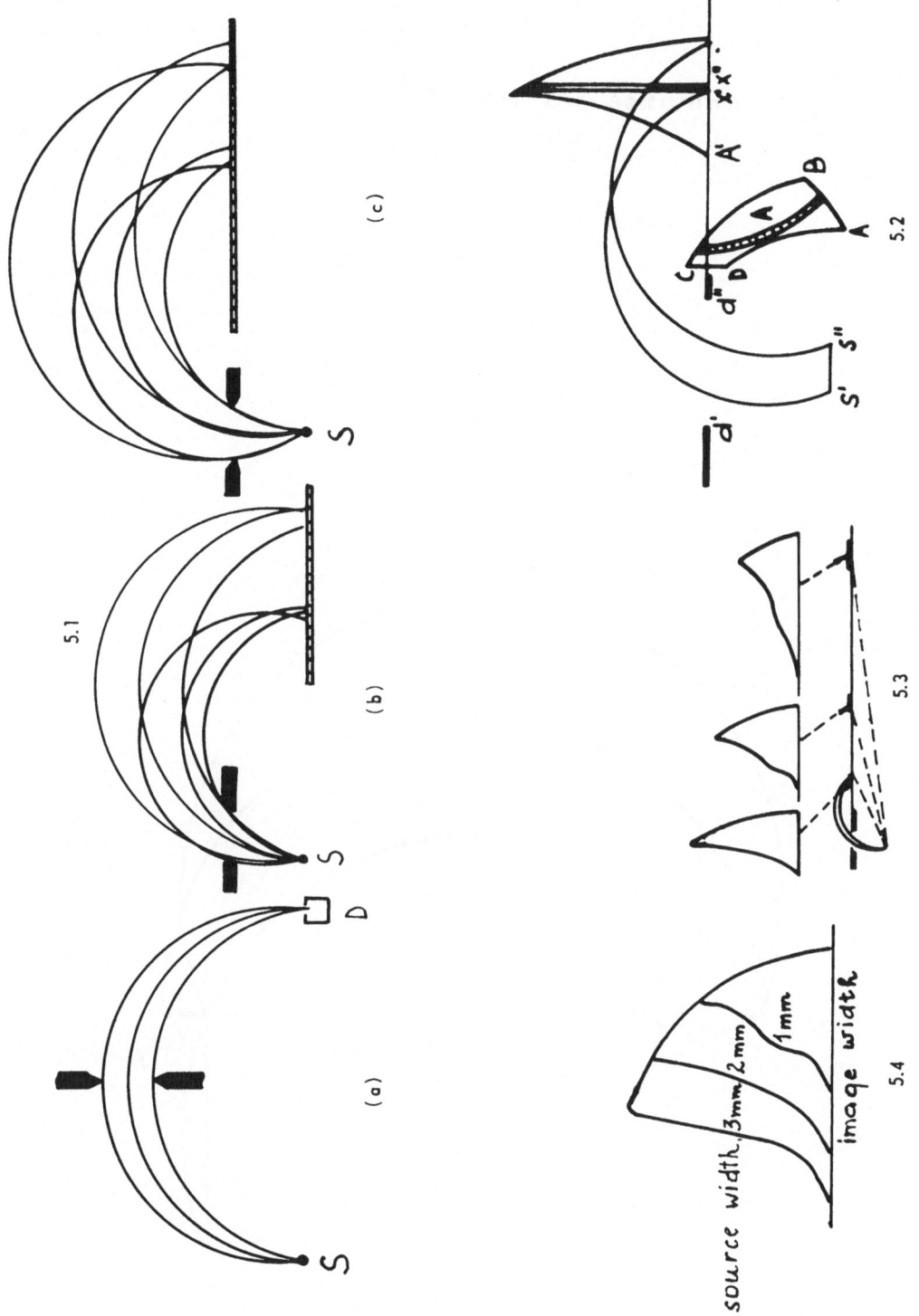

5.1

(a)

(b)

(c)

5.2

5.3

5.4

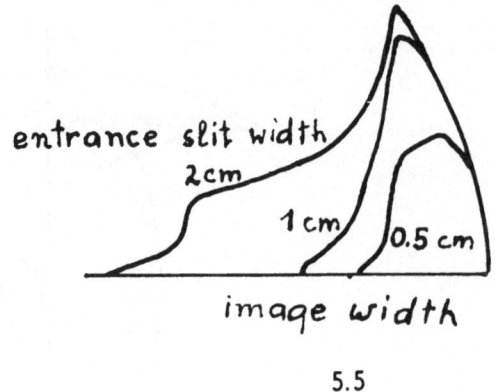

entrance slit width
2cm

1 cm

0.5 cm

image width

5.5

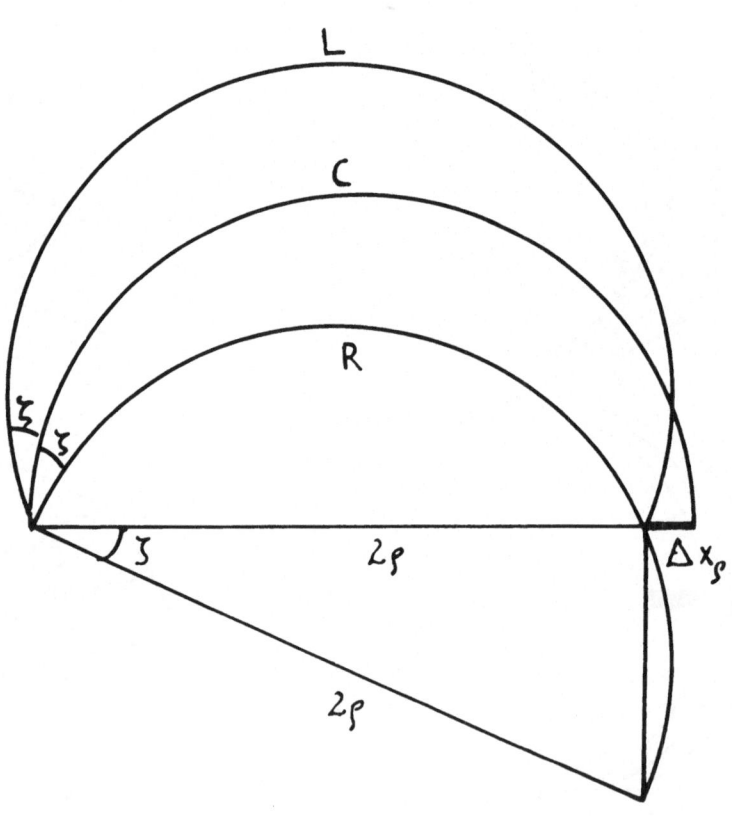

L

C

R

ζ ζ

ζ

2ρ

Δx_ρ

2ρ

5.6

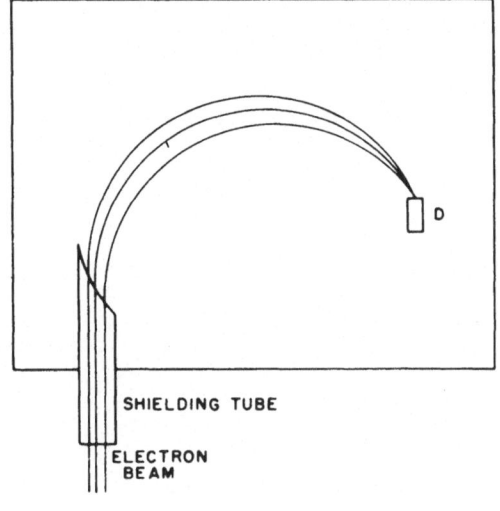

SHIELDING TUBE

ELECTRON
BEAM

5.7

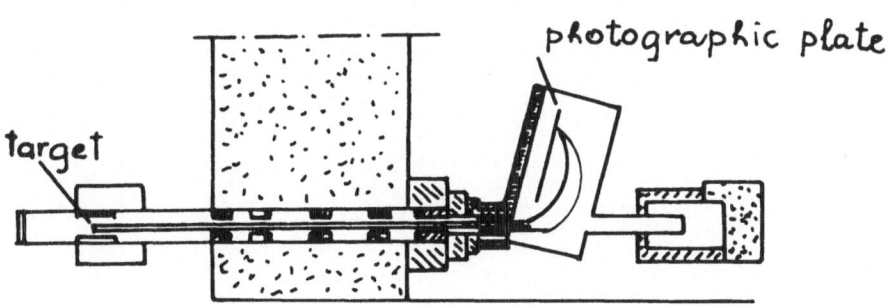

target

photographic plate

5.8

6. $\pi\sqrt{2}$ SPECTROMETER

The $\pi\sqrt{2}$ spectrometer has probably been the most usefull type of spectrometer since early fifties. Many laboratories have built one or two of them and the design was constantly improving.

The main problem in the development of this type of spectrometer has been the design of magnets which would produce the required field form with high accuracy. The latest ironfree (29-31) $\pi\sqrt{2}$ spectrometers have the fields very close to optimum.

Iron-core and iron free magnets were almost equally in favor, We shall consider them in turn after discussing the aberrations.

Discussion of second order solution – General solutions of the second order, (4.71) and (4.72), become for $\alpha_1 = -0.5$

$$\eta_o = -h + \frac{1}{3}(1 - 8\,\alpha_2)H_o^2 + (\frac{8}{3}\,\alpha_2 - 1)T_o^2$$
$$+ \frac{2}{3}(1 - 2\,\alpha_2)\,h^2 + (\frac{4}{3}\,\alpha_2 - 1)\,t^2 \tag{6.1}$$

$$\tau_o = -t + 2\,(\frac{8}{3}\,\alpha_2 - 1)H_o T_o + \frac{8}{3}\,\alpha_2\,th \tag{6.2}$$

We shall first analyze the contributions of source dimensions to the radial aberrations.

The first term represents the width of the source h in first order. It can contribute substantially to the line width. In high resolution work, h is often of the order of 10^{-3} .

The second order term in h contributes much less than the first order term and can be neglected.

The height of the souce t contributes to radial aberrations a term of second order. This indicates that rectangular sources can be used, the height being an order of magnitude larger than the width. There is also a possibility to cancel the contributions of t^2 term by proper choice of h term, so that

$$-h + (\frac{4}{3}\,\alpha_2 - 1)\,t^2 = 0 \tag{6.3}$$

The central line of the source should then follow the curve described by (3).
We should add, however, that in most cases this possibility is not used.

The remaining two terms represent the aberrations due to radial
and axial opening of the beam. They would be the only terms present in the
case of point source, which would have the base width of the image b given
by

$$b = pH_o^2 + qT_o^2 \tag{6.4}$$

where

$$p = \frac{1}{3} (1 - 8 \; \alpha_2) \qquad q = (\frac{8}{3} \; \alpha_2 - 1) \tag{6.5}$$

The coefficient α_2 should be chosen to optimise the transmission
for the given base width b. Using the Lagrange method of optimisation we
define a function U

$$U = b + \lambda \Omega$$

$$= pH_o^2 + qT_o^2 + \lambda \cdot \frac{1}{4 \pi} H_o T_o$$

The partial derivations are

$$\frac{\partial U}{\partial H_o} = 2pH_o + \frac{\lambda}{4\pi} T_o \tag{6.6}$$

$$\frac{\partial U}{\partial T_o} = 2qH_o + \frac{\lambda}{4\pi} H_o \tag{6.7}$$

Putting them equal to zero, and multiplying by T_o and H_o respectively, (6) and (7) give

$$2pH_o^2 = - \lambda \Omega \tag{6.8}$$

$$2gT_o^2 = - \lambda \Omega \tag{6.9}$$

Adding (8) and (9) we obtain

$$2(pH_o^2 + qT_o^2) = - 2\lambda \Omega$$

or

$$b = \lambda \Omega \tag{6.10}$$

so that (8) and (9) become

$$H_o^2 = \frac{b}{2p} \qquad\qquad T_o^2 = \frac{b}{2q} \qquad\qquad (6.11)$$

Using these optimum values of H_o and T_o, transmission becomes

$$\Omega = \frac{b}{4\pi} \cdot \frac{1}{\sqrt{pq}} \qquad\qquad (6.12)$$

The transmission is maximum for the minimum values of pq, which has the explicit form

$$pq = \frac{1}{3}(1 - 8\,\alpha_2)(\frac{8}{3}\,\alpha_2 - 1)$$

$$= \frac{64}{9}\,\alpha_2^2 - \frac{32}{9}\,\alpha_2 - 1 \qquad\qquad (6.13)$$

The quantity pq is equal to zero, and transmission maximum for

$$\alpha_2 = \frac{1}{8} \quad\text{and}\quad \alpha_2 = \frac{3}{8}$$

On the other hand pq is maximum and transmission minimum for $\alpha_2 = 1/4$ which can be seen by taking the derivative of (1/3) and making it equal to zero.

Since the formula for transmission which was used implies rectangular entrance baffle, the conclusions are valid only for that case. Rosenblum (31) finds that when circular entrance baffle is used, $\alpha_2 = 1/4$ gives the best transmission for the given resolution.

Only three values of the coefficient α_2 were used in designing beta spectrometers, and they are

3/8, 1/8, 1/4.

We shall consider each one in turn.

$\alpha_2 = 3/8$ - This choice of α_2 corresponds to high aperture type, since the term containing T_o^2 vanishes. It has the following advantages over the wide aperture type:

- As it was already noted, in high aperture spectrometers the term $H_o T_o$ in (2) vanishes. The contributions of the radial and axial opening to the height of the image are of the order higher than the second.

- The high aperture field is close to $1/r$ field. Within such a field the choice of central orbit is not unique and a whole range of momenta can be focused simultaneously in an extended focal plane.

- The coefficient of t^2 term is 66% greater in the case of $\alpha_2 = 1/8$ than for $\alpha_2 = 3/8$, which means that in high aperture spectrometers 30% higher sources can be used.

From the geometrical point of view, the high aperture type is suitable for the iron-free spectrometers with the coils of current sheet type.

$\alpha_2 = 1/8$ - The wide aperture type offers an economical advantage in the case of iron spectrometers, because a flat beam allows the use of relatively smaller pole distance. When the ratio of pole distance to pole radius is smaller, fringing effects are less pronouced, and it should be less difficult to produce a given field geometry.

$\alpha_2 = 1/4$ - In this case circular baffle is used. This choice of α_2 coefficient was practically limited to iron spectrometers, where it offers an advantage from the constructional point of view. The pole pieces have simple conical shape, which is much easier to machine, than in the case of $\alpha_2 = 3/8$ and $1/8$.

Higher order calculation - The central problem in the design of a spectrometer represents the successive fitting of coefficients in the series expansion of the magnetic field. In the iron spectrometers coefficients are fitted usually up to the second and rarely up to the third order. More precision can be achieved with iron-free spectrometers, where so far coefficients up to the sixth were fitted. The coefficients higher than second represent new degrees of freedom available for the reduction of aberrations. They are chosen to minimize the important higher order terms in H_o and T_o.

The choice of higher order coefficients depend of course on the choice of α_2. If, for instance, α_2 is equal to $3/8$, axial opening T_o can be much larger than H_o. The dominant higher order terms are those containing higher powers of T_o and they put the upper limit to axial opening which can be used for a given resolution. Higher powers of H_o, which is much smaller, can be neglected, since they appear in first order and are much smaller than T_o. The important higher order terms are, therefore, $H_o T_o^2$, T_o^4, $H_o T_o^4$, T_o^6... These terms can be annulated by proper choice of coefficients α_3, α_4, α_5, α_6, ..., each coefficient annulating one term.

Higher order solutions of equations are successively obtained as it was done for second order solution, which was found by starting from first order solutions.

A given magnet design usually fits a limited number of field coefficients. Some higher order terms of the produced field might deviate appreciably from the theretical values. These difficulties can be overcome now using computers, especially for iron-free spectrometers. The field is calculated in the whole volume where trajectories pass and then by computing a large number of orbits, isoaberration curves can be obtained and resolution-transmission ratios calculated. The coils parameters can then be varied to optimise the resolution-transmission ratio. This has been done by the Uppsala group (Besev, Castman, Norberg, Olsen, ans Pettersson, ref. 33) for the spectrometer described in ref. 29. Their results are illustrated in Figs. 1-2. Fig. 1, shows the theoretical optimum field and two fields for iron-free magnets, denoted by 30–46 and 30–47, where numbers denote the numbers of turns in inner and outer coils. It appears from Fig. 1 that 30–46 is a better fit to the optimum field, but Fig. 2 shows that 30–47 has better resolution-transmission characteristics.

The corresponding field coefficients are given below.

Field	α_1	α_2	α_3	α_4	α_5	α_6
Optimum	− 0.5000	0.3750	− 0.2986	0.2400	− 0.2017	0.5768
30–46	− 0.5001	0.3828	− 0.4629	−6.2716	− 5.1390	6.6073
30–47	− 0.4986	0.3616	− 0.4704	−6.1648	− 5.4456	4.8058

It can be seen that the field coefficients of two possible magnets differ appreciably from the theoretical ones, beginning with α_4. Then computer calculated performances represent the best way to find the optimum magnet parameters.

Iron magnets – There is a choice of two geometries for the iron--core double-focusing spectrometer. In one case the coil is in the center, while the other is called insideout type, having the coils surrounding the working gap of spectrometer. The coil-in-the-center type was adopted by Siegbahn,

Swartholm and Hedgran (3) in their pioneering design of first large double-focusing spectrometer shown in Fig.3. The first spectrometer of the inside-out
type was designed by Bartlett and Bainbridge (10), (Fig.4).

The advantages of central coil type are

- Economy of copper and iron

- Accessibility to the chamber

The disadvantage is a difficult handling of edge effects.

The inside-out type costs more and the accessibility is reduced,
but the flux leakage is eliminated and the edge effects reduced.

The experience with about twenty iron-core double-focusing spectrometers of both types built so far (1-15), some of them being copies of the same
design, shows that it is hardly possible to obtain the exact field-form by the
theoretical design of the pole-pieces profiles. Sometimes an up-down assym-
metry is found. The field shape may vary with the field intensity, especially
at very low energies. Often the large space, intended for high transmission is
of no use, because the field deteriorates so much at relatively larger distances
from the equilibrium orbit, that at larger apertures the line width increases
prohibitively.

The field form can be corrected by iron shimming or coils. A
flexible shimming system was devised by Samoilov (6). Iron rings are distributed outside and inside the working gaps, as shown in Fig.5. Some of them
are close to the iron in the conventional manner, but he finds it useful to have
some of the rings close to the median plane. The field shape in the midian
plane remains constant within 0.15% for the intensities from 6 to 780 gauss.

A coil corrector for central coil type is shown in Fig.6. Kovrigin,
Kolesnikov and Latishev (8) used the pole profiles previously studied by
Grigorieff and Zolotavin, adding two flat coils, parallel to the median plane.
The field form is then fitted within 0.03%.

In the inside-out type, the excitation coil can be divided in several
sections and the current varied in them, to correct the field. Fig.7 shows such
a spectrometer designed by Bartlett, Ristinen and Bird (14). Besides seven
outer coils they found it necessary to place three inner coils.

Iron-free spectrometers - Two kinds of geometry were used in the design of coils for iron-free spectrometers. One approach is to start with pairs of current loops, symmetric with respect to median plane containing the orbit circle. The number of pairs varies usually between two to four (See Fig. 8). Larger numbers of loops give more parameters to fitt the field form, but requires more care in mechanical construction.

Another approach uses current sheets. The number of solenoids varies between two to four (See Fig. 9).

In both cases finite size of coils has to be taken into account, once the parameters of ideal loops and sheets are found, which give the desired field form.

First iron-free $\pi\sqrt{2}$ spectrometer was designed by Siegbahn (16). The desired field is produced in the region between two coaxial long coils. Five parameters are available to fit the field form: Two radii, two lengths and current ratio of coils. The outer coil gives the main contribution to the field, while the inner one serves as a correction. The currents in the coils are flowing in opposite directions. The radius of the optical circle was $r_o = 30$ cm.

An identical spectrometer was made by de Vries and Wapstra (21) in Amsterdam. The same geometry, scaled up to $r_o = 50$ cm was used by Mladjenović for a spectrometer constructed in Belgrade (19) and another in Cairo. Two spectrometers of this type, with $r_o = 50$ cm and somewhat changed parameters, have been built by Siegbahn and collaborators, one in Berkeley (25) and another in Uppsala (26-27).

The two coil spectrometers are relatively simple to construct and align, while the access to source and detector is not so good. No higher field coefficients than α_2 were fitted and in all cases α_2 was equal to 3/8.

The latest development of current sheet type of $\pi\sqrt{2}$ spectrometers is again due to Seigbahn and his collaborators (29). They designed a spectrometer with two pairs of coils, symmetrical with respect to median plane (Fig. 9). In that way two more parameters are obtained, and with help of computers the optimum coil parameters were found as it was discussed in the preceding section.

Current loop type of $\pi\sqrt{2}$ spectrometers was first built by Moussa and Bellicard (18). Four pairs of coils, all having radii larger than the radius of the optic circle, were designed to fit the desired field up to α_2, which was also taken to be 3/8. The same coil geometry, with somewhat scaled up dimensions were used for construction of a spectrometer in Nashville (22) and another in East Lansing.

Baranov (23) built a small spectrometer ($r_o = 10$ cm) with two pairs of coils fitting the field coefficients calculated by Pavinskii.

An unprecedented effort, in theoretical as well as experimental work, was made by Chalk River group, in construction of a $r_o = 100$ cm iron--free $\pi\sqrt{2}$ spectrometer, the largest beta spectrometer so far built. We will only briefly mention the main points in the magnet construction and testing:

- Lee-Whiting and Taylor calculated the optimum field coefficients up to α_6.

- They also developed theoretical method to calculate corresponding coils, which was applied to obtain more precise parameters for a three pairs of coils geometry, found by Wolfson, Graham and Ewan to produce roughly a $1/r$ field. These coils were fitting the theoretical field up to α_5. Lee-Whiting also calculated the manufacturing tolerances on coil parameters.

- Graham, Ewan and Geiger (24) designed and built the spectrometer which has a constancy in the calibration ratio, gauss cm per ampere, of 1 part in 10^5 for electron energies up to 4 MeV. The differences in mean radii and deviation of circularity of coils which had diameter from 0.9 to 4 meters were less than 2 parts in 10^4. The coil distances were constant within 5 parts in 10^4. The cooling system was designed to produce temperature constancy of the order of 1^oC.

- Lee-Whiting returned to actual coils and calculated the orbits and isoaberration curves. Geiger and Graham (34) measured the isoaberration contours. Theoretical and experimental isoaberration curves are shown in Fig.10. The entrance baffles were cut according to isoaberration curves for optimum performance.

Scaled down versions of Chalk River $\pi\sqrt{2}$ spectrometer were built in several laboratories: Ottawa (28), Pasadena, Grenoble (35), Oak Ridge (30) and Idaho Falls (31).

Two-fold focusing spectrometer – In special cases when background produced by outside sources is important, it can be reduced by an arrangement developed by Shestopalova (36). It belongs to a class of multiple focusing spectrometers, which have been built by Dzelepov and his collaborators in Leningrad.

The detector is transparent to electrons, having very thin double windows, so that the beam after passing through the focus makes another 255° and hits the second detector. Placing the source above the median plane, the first detector is located below and the second again above the median plane. Both detectors are in coincidence. If it is necessary, two detectors are lined up in the second focus and triple coincidences counted.

This arrangement requires wide beam geometry. Scattering in the first detector produces less problems when higher energies are measured. Resolution is determined by the slit in front of the first plane, the second focus having a larger slit to accept scattered electrons.

References Ch.6

1. F.Kurie, J.S.Osaba and L.S.Slack, RSI 19, 771 (1948)

2. F.B.Shull and D.M.Denison, Phys. Rev. 71, 681 (1947)

3. A.Hedgran, K.Siegbahn and N.Swartholm, Proc. Phys. Soc. Lond. Ser. A 63, 960 (1950)

4. M.W.Jones, H.Watterman, D.Mc Askill and C.D.Cox, Canad. J.Phys. 31, 225 (1953)

5. E.P.Grigoriev, Yu.S.Egorov, A.V.Zolotavin, V.O.Sergeev and M.S.Sovtsov, Izv. Akad. Nauk SSSR, Ser.Fiz. 29, 721 (1965)

6. P.S.Samoilov, Prib. Tekn. Eksp. No.6,33 (1959)

7. J.Katoh, N.Nozava, Y.Yoshizawa, Y.Koh, Journ. Phys.Soc.Japan 15, 2140 (1960)

8. C.D.Kovrigin, N.V.Kolesnikov, and G.D.Latishev, Prib. Tekn. Eksp. No 2, 19 (1961)

9. I.Rezanka, A.Spalek, J.Kuklik and J.Frana, Cesk. Cas.Fys. 17, No 1, 43 (1967)

10. A.A.Bartlett and K.T.Bainbridge, RSI 22, 517 (1951)

11. P.H.Stoker, Ong Ping Hok, E.F.de Haan and G.J.Sizoo, Physica 20, 337 (1954)

12. E.Arbman and N.Swartholm, Ark. Fys. 10, 1 (1955)

13. O.Huber, L.Schellenberg and H.Wild, Helv, Phys.Acta 33, 534 (1960)

14. A.A.Bartlett, R.A.Ristinen and R.P.Bird, NIM 17, 188 (1962)

15. Z.Playner, L.Maly, Ces.Cas. Fys. B 10, 476 (1960)

16. K.Siegbahn, Physica 18, 1043 (1952)

17. K.Siegbahn and K.Edwardson, Nucl. Phys. 1, 137 (1956)

18. A.Moussa and J.B.Bellicard, J.Phys. Rad. 17, 532 (1956)

19. M.Mladjenović, Proc. Rehovoth Conf. 1957, North Holland 1958

20. J.B.Bellicard, Ann.Phys. 13, 419 (1957)

21. C.de Vries and A.H.Wapstra, NIM 8, 121 (1960)

22. G.L.Baird, J.C.Nall, S.K.Itaynes and J.H.Hamilton, NIM 16, 275 (1962)

23. V.F.Baranov, Prib. Tekn. Eksp. No 3, 15 (1958)

24. R.L.Graham, G.T.Ewan and J.S.Geiger, NIM 9, 245 (1960)

25. K.Siegbahn, C.Nardling and J.M.Hollander, UCRL - 10023 (1962)

26. K.Siegbahn and C.Nordling, Ark. Fys. 22, 436 (1962)

27. A.Fahlman, S.Hadström, K.Namrin, R.Nordberg, C.Nordling and K. Siegbahn, Ark. Fys. 31, 479 (1966)

28. J.L.Wolfson, W.J.King and J.J.Park, Can.J.Phys. 41, 1489 (1963)

29. R.Nordberg, J.Hedman, P.F.Heden and C.Nordling, Ark.Fys. 37, 489 (1968)

30. C.E.Bemis, Jr., ORNL - 4306

31. R.G.Helmer, in Nucl.Tech. Branch, Progr. Rep. IN-1218 (1968)

32. E.S.Rosenblum, Phys. Rev. 72, 731 (1947)

33. C.Besev, B.Castman, R.Nordling, B.Olsen and G. Pettersson, NIM 62, 125 (1968)

34. J.S.Geiger and R.L.Graham, NIM 24, 81 (1963)

35. A.Baudri and A.Moussa, NIM 72, 167 (1969)

36. Y.Shestopalova and B.S.Dzhelepov, Izv.Akad.Nauk SSSR, Ser.Fiz.25, 1302 (1961)

Text to Figures. Ch. 6

Fig. 6.1.   The theoretical optimum field (T) and two calculated fields (30-46 and 30-47).

Fig. 6.2.   Resolution-transmission characteristics for calculated fields.

Fig. 6.3.   First large double-focusing spectrometer (Siegbahn, Swartholm, Hedgran).

Fig. 6.4.   First DF spectrometer of inside-out type. (Bartlett, Bainbridge).

Fig. 6.5.   The shimming system designed by Samoilov (marked by number 1.).

Fig. 6.6.   A corrector for central coil (marked by number 2). (Kovrigin, Kolesnikov, Latishev).

Fig. 6.7.   The correctors for inside ← out type. (Bartlett, Ristinen, Bird).

Fig. 6.8.   Two geometries using current - loops approximation. (a) Chalk River instrument. (b) later simplification of chalk River geometry (ref. 28, 30, 31, 35).

Fig. 6.9.   Two geometries using current - sheet approximation. (a) Early Siegbahn´s geometry (ref. 16, 21, 25, 26, 27), (b) later Siegbahn´s geometry (ref. 29).

Fig. 6.10. The isoaberration curves for Chalk River spectrometer

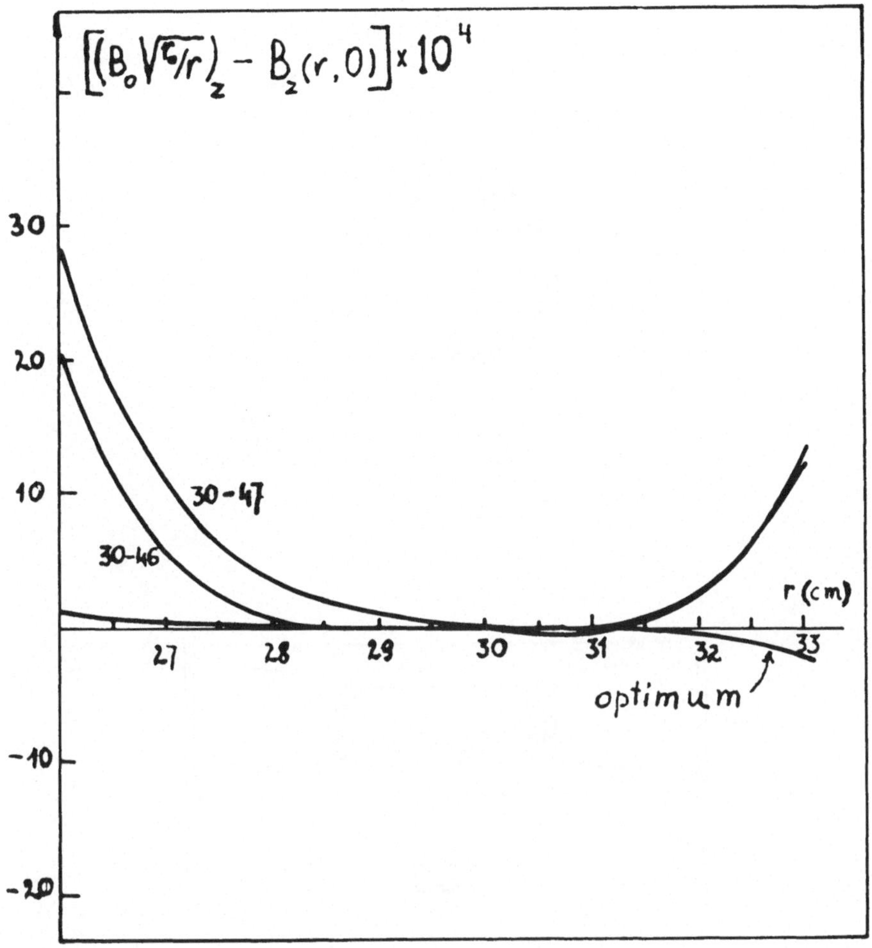

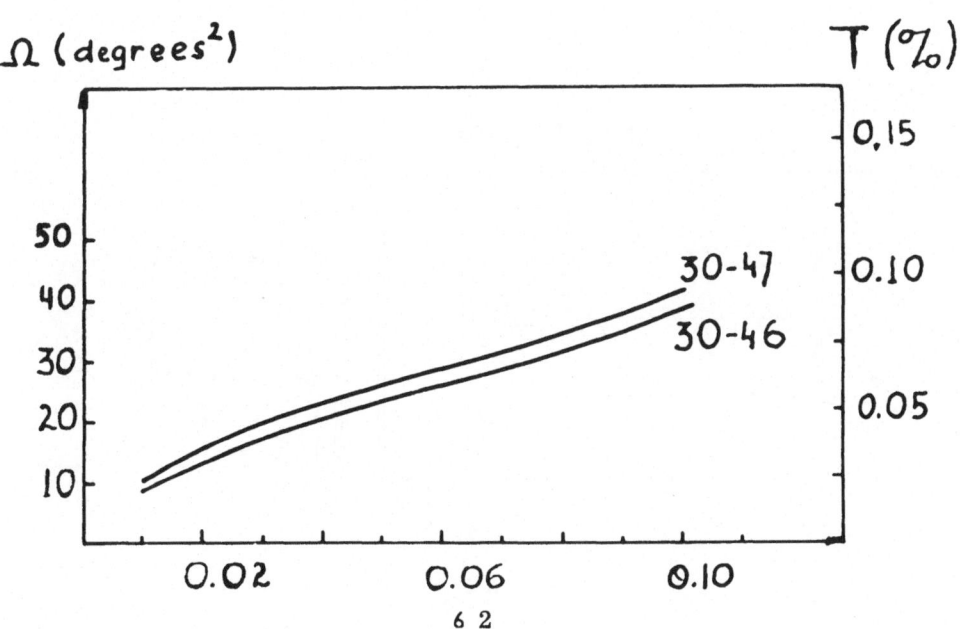

[(B0√(r0/r))z − Bz(r,0)] x 10^4
30
20
10
-10
-20
30-47
30-46
r (cm)
27
28
29
30
31
32
33
optimum

6.1

Ω (degrees²)
T (%)
0.15
50
40
30
20
10
0.10
0.05
30-47
30-46
0.02
0.06
0.10

6 2

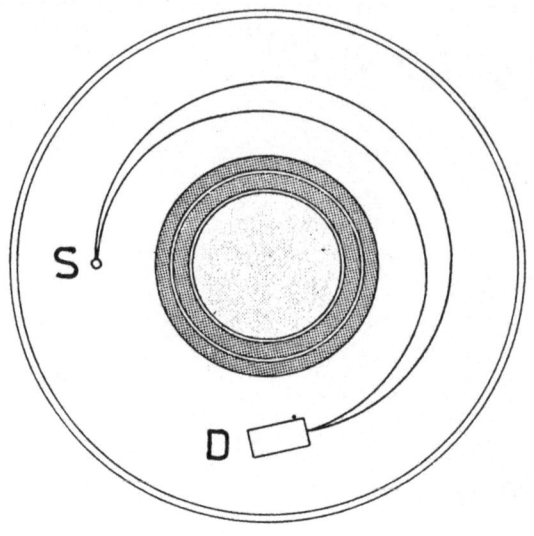

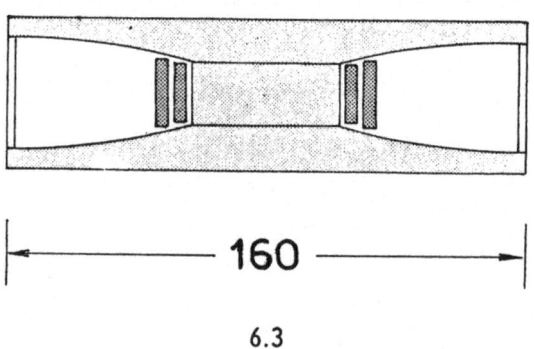

160

6.3

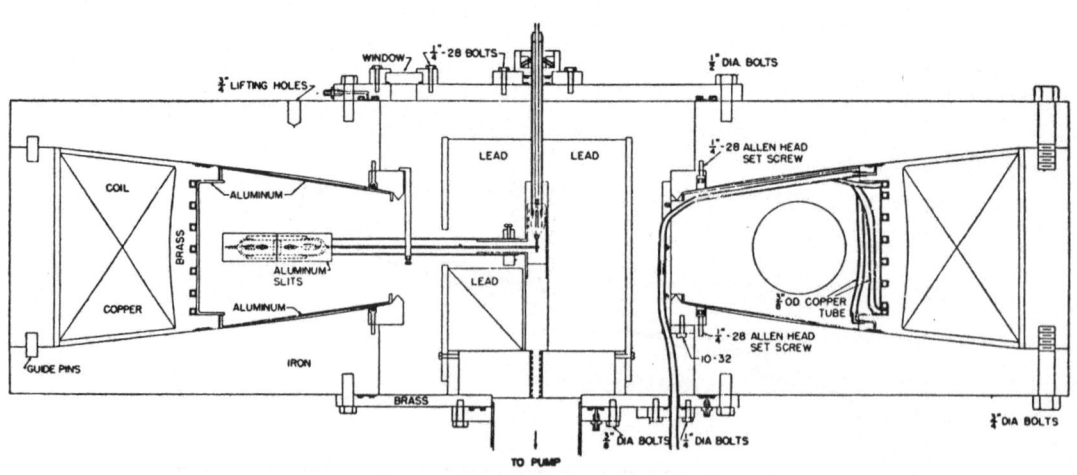

6.4

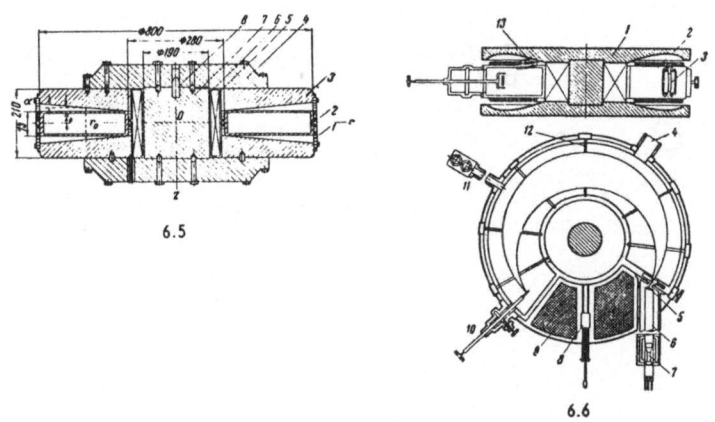

6.5

6.6

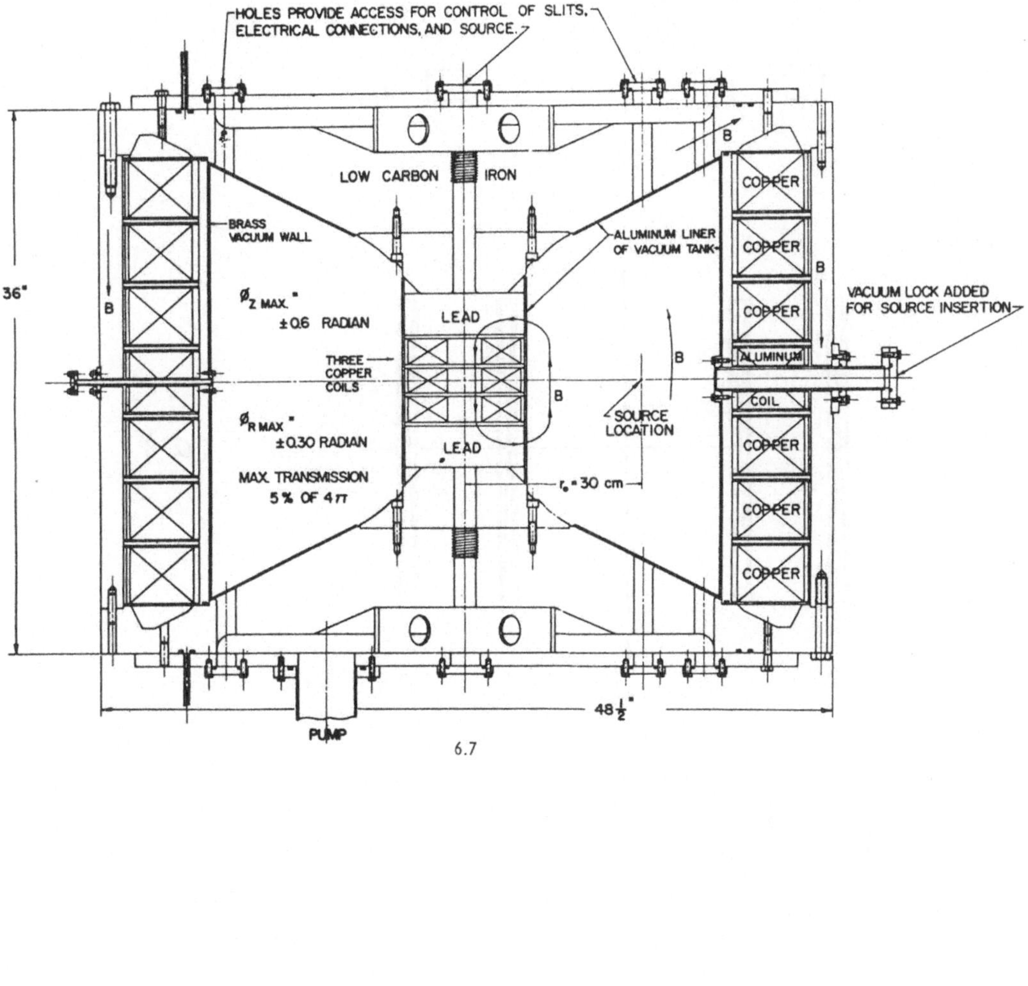

HOLES PROVIDE ACCESS FOR CONTROL OF SLITS,
ELECTRICAL CONNECTIONS, AND SOURCE.

LOW CARBON IRON

COPPER

COPPER

COPPER

BRASS
VACUUM WALL

ALUMINUM LINER
OF VACUUM TANK

$\phi_{Z\ MAX.}$ =
± 0.6 RADIAN

LEAD

ALUMINUM

VACUUM LOCK ADDED
FOR SOURCE INSERTION

THREE
COPPER
COILS

COIL

$\phi_{R\ MAX.}$ =
± 0.30 RADIAN

SOURCE
LOCATION

COPPER

MAX. TRANSMISSION
5 % OF 4π

LEAD

COPPER

r_o = 30 cm

COPPER

36"

48 ½"

PUMP

6.7

(a)

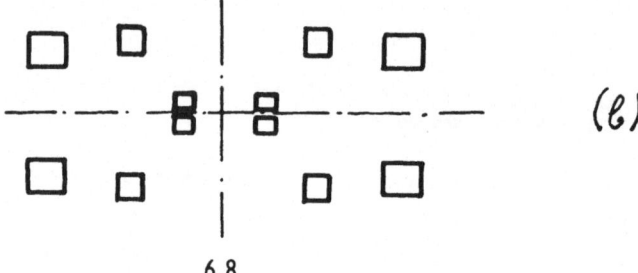

(b)

6.8

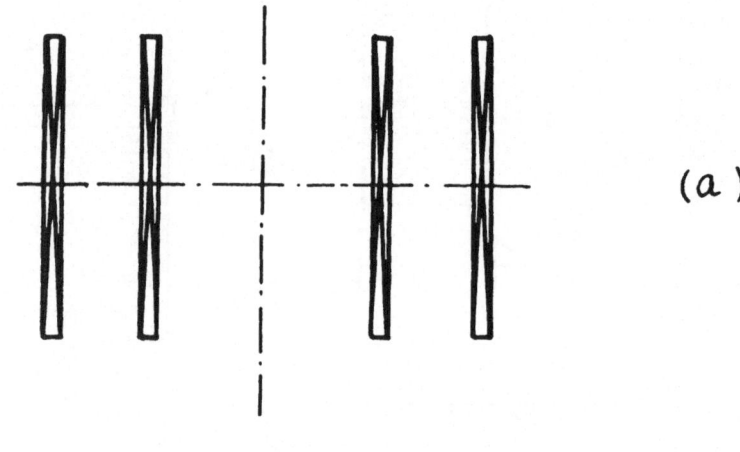

(a)

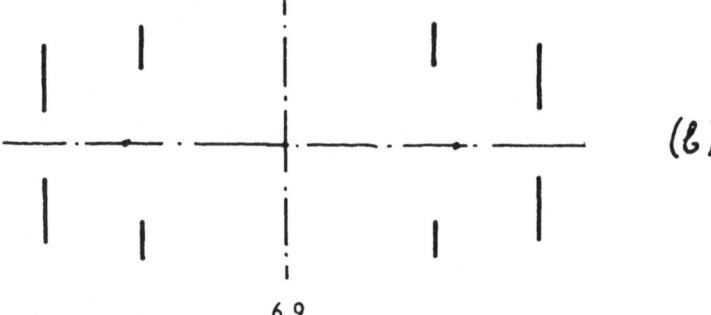

(b)

6.9

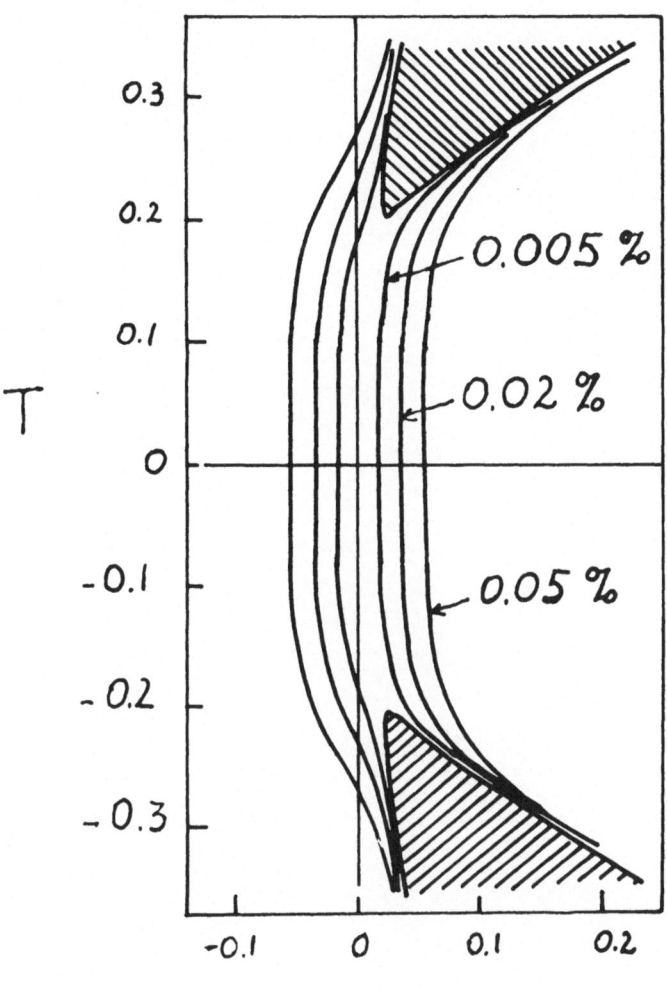

6.10

7. $\pi\sqrt{10}$ SPECTROMETER

It was shown in Ch.4 that by increasing the radial focusing angle of a double-focusing spectrometer, the dispersion and the luminosity also increases. Starting from $\pi\sqrt{2}$, in which radial and axial focusing angles are equal (n = 1), next double focusing is obtained for n = 2, where radial focusing angle is twice larger than the axial one. In general, there are n-1 intermediate axial focalisations, for a single axial one. Lee-Whiting has shown (1) that provided H_o^4 is the first uncancelled power of H_o in wide beam radial aberration formula, the luminosity increases roughly as cube of n at very high resolutions. On the other hand the relation (4.70) shows that dispersion increases as the square of n. Higher values of n represent therefore certain advantages which have to be weighted against serious inconvenience that the beam hits the detector and the source before focalisation.

Another important question is to see, whether aberrations increase with n. This was also analyzed by Lee-Whiting in his fundamental paper on high-dispersion spectrometers, and he shows that no second order aberration increases appreciably as n increases. We shall illustrate these conclusions by considering the coefficient of T_o^2 in the general expression for second order solutions (4.71). Since wide beam is generally preferable, when beam is hitting the detector, we put the coefficient of H_o^2 in (4.71) equal to zero, and using (4.59) obtain

$$\alpha_2 = (n^2 - 3/4)/(n^2 - 1) \tag{7.1}$$

With this value of α_2 the coefficient of T_o^2 in (4.71) becomes

$$-2n^2 (4n^2 - 1)^{-1} \tag{7.2}$$

It changes monotonically from $-2/3$ to $-1/2$ as n increases from one to infinity. The coefficients of other second and third order terms behave similarly for both, wide and high aperture types.

Another two conclusions from Lee-Whiting analysis should also be mentioned:

- Odd values of n are preferable to even ones, because then the ratio of beam width to the obstacles (source and detector) size is more favorable.

- If the source emits continuous spectrum, some of the electrons enter the detector on their first encounter with it. This kind of background increases with n.

$\pi\sqrt{10}$ spectrometer in Moscow - The first and the only $\pi\sqrt{10}$ spectrometer built so far was designed and constructed by S.A. Baranov and collaborators (2) in Kurtchatov's Institute. Only brief informations on the spectrometer were published. The basic parameters of the spectrometer are the following:

$n = 3$

$D = 20$

Radial focusing angle $= 570^{\circ}$

Radius of equilibrium orbit $r_o = 40$ cm

Field shape $B = B_o (1 - 0.9\eta + 0.825 \; \eta^2 - 0.767 \; \eta^3)$

Although it is not stated explicitly, one can gather from the note, that the magnet is made with iron. Since the radial opening angle of pole-pieces increases with α_1, the fringing field effects become more serious and design in general more involved. An iron-free magnet would be a much more convenient choice, as stated also by authors themselves in the last sentence of the paper.

The beam is prevented to enter the detector before completing 570° by two diaphragmes. (Fig.1). One of them placed at 285° cuts the central part of the beam, and another placed in front of the source limits its radial opening angle. In such a way 2/3 of the horizontal opening of the beam is lost.

The authors give the following performances measured with a ^{137}Cs source:

Source dimensions: 1.5×40 mm^2

Counter slit dimensions: 2×50 mm^2

Solid angle: 3×10^{-4} of 4π

Half width: 0.038%

Geometric luminosity: 1.8×10^{-2} mm^2

References for Ch. 7

1. G. E. Lee-Whiting, Can. J. Phys. <u>35</u>, 570 (1957)

2. S. A. Baranov, R. M. Polevoi, Yu. A. Aliev and S. N. Belenkii,
   Prib. Tekn. Eksp. No. 6, 64 (1965)

118

Text to Figures. Ch. 7

Fig. 7.1.   The geometry of beam and baffles. Focusing beam is defined by central baffle (1); the potential background beam is cut by (2) and (3).

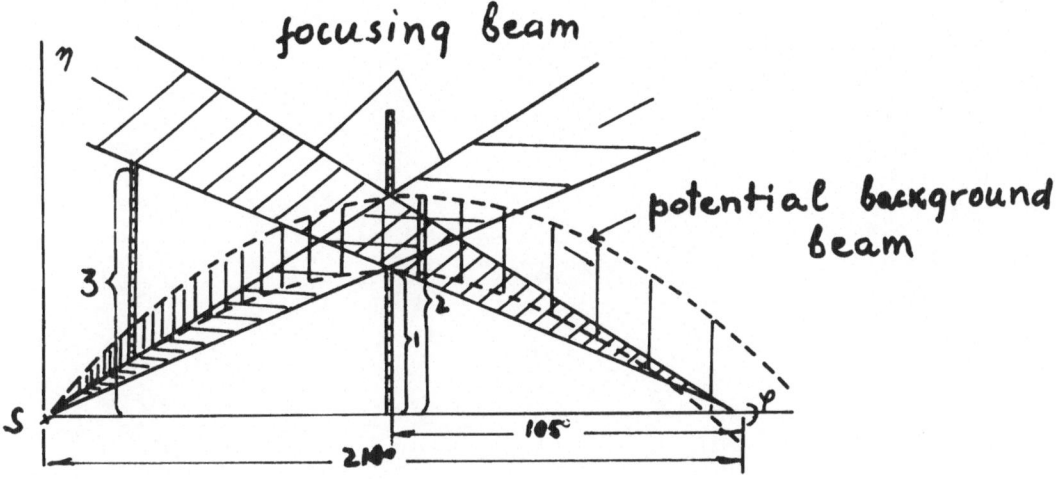

focusing beam

potential background beam

8. $(\pi/2)\sqrt{13}$ SPECTROMETER

The only spectrometer with half integer number of axial focalisations was built by Jahn, Daniel, et. al., (1-4) in Heidelberg. The idea of Daniel was that an axially extended focus offers new degrees of freedom, since η and θ do not have to remain constant. In a wide beam spectrometer, for instance, where the coefficient of H_o^2 term was put equal to zero, the contribution to the aberrations of the remaining T_o^2 term can be cancelled by curving the source in ϕ_o plane. The equation of the exit slit which is very simple for the point source can be found by starting from general second order solutions (4.71-72) and introducing the following conditions:

1. Focusing angle ϕ_o is given by the relation (4.61)

$$\phi_o = \phi_r = \pi/\sqrt{1+\alpha} \mp (2N - 1)\, \pi/2\sqrt{-\alpha_1} \qquad (8.1)$$

or

$$\sqrt{-\alpha_1/(1 + \alpha_1)} = N + 1/2 \quad \text{where N is integer} \qquad (8.2)$$

2. The coefficient of H_o^2 is put equal to zero, giving

$$\alpha_2 = - (3 + 7\,\alpha_1)/4 \qquad (8.3)$$

With these two conditions and the point suurce, (4.71) and (4.72) reduce to

$$\eta_f = - (3/4)\, T_o^2 \qquad (8.4)$$

$$\tau_f = (-1)^{N+1}\, T_o \qquad (8.5)$$

which gives the following equation of the extended focus

$$\eta_f = (3/4)\, \tau_f^2 \qquad (8.6)$$

The relation (5) shows that each axial emission engle has a corresponding height.

Daniel pointed out (1) that the dependence of resolution on some higher order terms can be eliminated. To begin with, the coefficients of higher order terms in H_o can be annulated by proper choice of corresponding field coefficients, α_3 for H_o^3, α_4 for H_o^4, etc. Since for symmetry reasons only even powers of T_o can appear in the relation for η_f, the only third order form, besides H_o^3 is $H_o T_o^2$. Daniel shows that this term can be eliminated by using another degree of freedom still left, the θ coordinate. Starting from θ_f in median plane, at the height τ_f, the third coordinate of the focus θ is displaced from θ_f by an amount proportional to τ_f^2. Passing to fourth order terms, T_o^4 can be added to eq. (4) as a correction of curvature, and contribution of $H_o^2 T_o^2$ term is reduced by a diamond shaped entrance baffle.

Daniel and Laslett (2) have determined the shape of exit slit by calculating the electron orbits with different H_o and T_o values. They assume that the shape of exit slit is given by the equations

$$\theta_f = \theta_1 + a_2 \, \tau_f^2 + a_4 \, \tau_f^4 + a_6 \, \tau_f^6 \tag{8.7}$$

$$-\eta_f = b_2 \, \tau_f^2 + b_4 \, \tau_f^4 + b_6 \, \tau_f^6 \tag{8.8}$$

and find the coefficients a and b. The resulting slit shapes in θ - t and $-$t planes are shown in Fig.1. They have also calculated that the contribution of aberrations to the resolution would be 1.10^{-6} for the solid angle $\Omega = 0.4\%$, (-7.3×10^{-6}) for $\Omega = 1.3\%$ and (-4.10^{-4}) for $\Omega = 2.7\%$. These are rather small aberrations. If one takes also into account that the dispersion of $(\pi/2)\sqrt{13}$ spectrometer is 6.5 and that high sources can be used, because the first order term in t vanishes from τ_f, one can expect excellent resolution-luminosity characteristics.

The basic parameters of the Heidelberg spectrometer are the following:

Magnet: Iron-free, 10 coils

Field shape: $B = B_o \, (1 - 0.692308 \, \eta + 0.461538 \, \eta^2 - 0.31656 \, \eta^3 + 0.22965 \, \eta^4)$

Radial focusing angle: 324.5^o

Dispersion: $D = 6.5$

Radius od Equilibrium orbit: $r_o = 30$ cm

Maximum energy: 4.5 MeV

The coil geometry (fig. 2) was first found using current-loop approximation and then corrections for finite size were made. The coils were made with a mechanical precision of 0.1 mm. Coils are cooled by water. In the symmetry plane, the deviations of the real field from the required theoretical field shape, are $(1-4)\cdot 10^{-4}$, as can be seen in Fig. 3.

The baffles defining the beam are shown in Fig. 4. As explained above, entrance baffle is diamond shaped, to reduce the fourth order aberration cross-term. The baffle placed in axial focus conveniently supresses the scattering and reduces the background. The detector slit has a parabolic shape, as required by equation (6).

Two G. M. counters are used, one of them 10 cm high for transmissions up to 0.7% and another 20 cm high for transmissions up to 1.5%.

The actual performance of the spectrometer can be judged from the experimentally found parameters (7) given in the Table I.

Table I

Source area, cm^2	R, %	T, %	L_g, cm^2
0.02 x 1	0.048	1	$2\cdot 10^{-4}$
0.02 x 1	0.03	0.6	$1.2\cdot 10^{-4}$
0.02 x 0.25	0.022	0.6	$3\cdot 10^{-5}$
0.02 x 0.25	0.013	0.16	$1.5\cdot 10^{-6}$

An inspection of the above table shows that the theoretically predicted performances were not attained, but the performances are still very good and compare favorably with those of $\pi\sqrt{2}$ spectrometers. The performances can probably be improved by more precise computer calculation of coils and corresponding isoaberration curves.

It should be mentioned that this type of spectrometer has the following disadvantages:

    - large detectors are needed

    - source-detector distance is rather small

    - multiple counter array is difficult to install.

References Ch. 8

1. H.Daniel, RSI 31, 249 (1960)

2. H.Daniel and L.J.Laslett, NIM 10, 48 (1961)

3. H.Daniel and P.Jahn, NIM 14, 353 (1961)

4. H.Daniel, P.Jahn, M.Kuntze and G.Spannagel, NIM 35, 171 (1965)

Text to Figures. Ch. 8

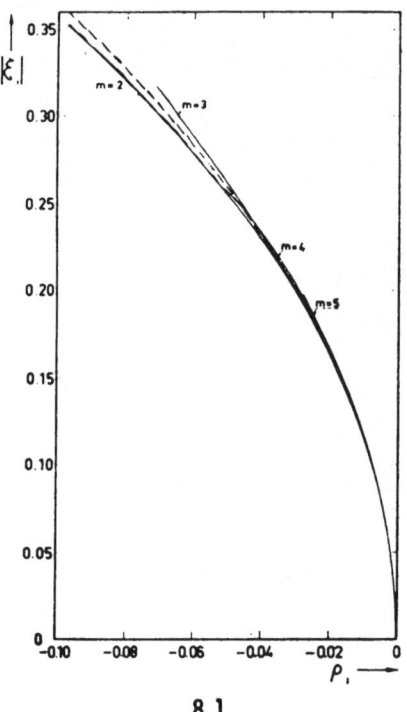

8.1

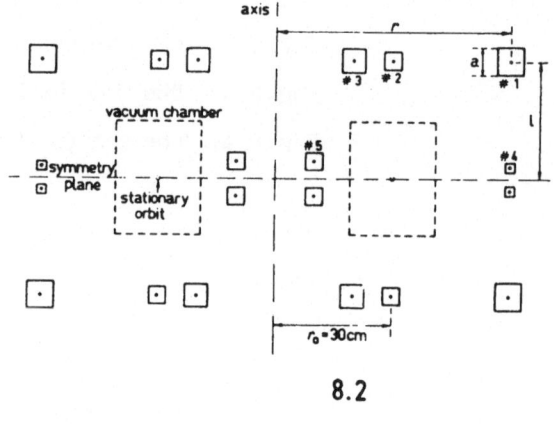

8.2

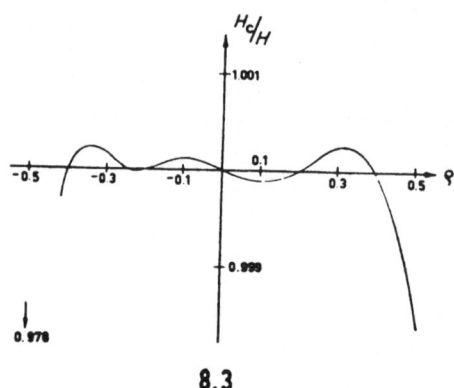

8.3

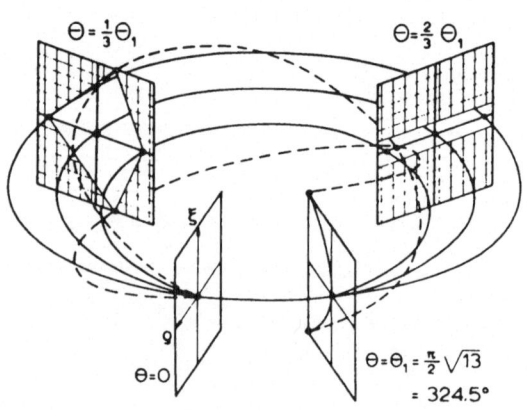

8.4

9. PRISMATIC (SECTOR) SPECTROMETERS

The advantage of sector magnetic spectrometers is that the source and the detector are in the field free region, allowing an easy access and use of auxilliary apparatus. They are convenient for analysis of beams produced in accelerators and ion sources. In beta spectroscopy, sectors are used for coincidence and angular correlations work.

It is not easy to make a precise theoretical design of sector spectrometers, due to the fringing flux effects. In most cases the fringing field cannot be accurately calculated, but has to be measured once the magnet is made. The theoretical design needs therefore the empirical corrections.

The disturbance produced by the fringing field will be the smallest for the case of a narrow beam, confined to median plane and entering perpendicularly to the sector boundary. The effect is then practically the same as if the effective boundary of the field were outside of the sector boundary at a distance of the order of the gap width. The disturbance increases when the beam spreads radially, remaining in the median plane, and becomes more serious for beams with axial openings. This rough, qualitative description of fringing field disturbances explains why the sectors were mostly used for analyses of narrow beam, less for flat but wide beams confined to median plane, and even less for beams having large axial apertures.

The mass-spectroscopy and accelerator beam guiding and analysis often deal with relatively narrow beams, or have to tolerante them, as sectors are the only spectrometers they can use. For those reasons in these fields by far the largest contributions have been made to the development of sector spectrometers. Since that work is fully covered in books and papers cited at the end of the chapter (1-3), we shall only give some elementary derivations and briefly describe some of the typical sector beta spectrometers. Additional information on sectors and fringing field effects are given in chapters on toroidal spectrometers and optical analogy spectrometers.

The study of sectors, since the first paper of Barber (4), can be roughly divided into following categories:

a) Uniform field, straight boundaries, fringing field effects neglected, motion in median plane. Basic formulae developed by Barber (4) Stephens (5), Herzog (6) and Cartan (7).

b) Uniform field, curved boundaries, fringing fields neglected, motion in median plane. Higher order radial focusing was studied by Kerwin (8), Dempsey (9) and Hintenberger (10).

c) Uniform field, straight boundaries, double-focusing produced by the fringing field. Gross (11) examined double-focusing in a single sector, and Camac (12) proposed a combination of two sectors.

d) Cylindrically symmetric field, radially decreasing as r^{-n}, straight boundaries, dobule-focusing produced by the sector field. Field with $n = -0,5$ was studied by Svartholm (13) and Rosenblum (14). Judd (15) considered the general case of the field r^{-n}, and later, with Bludman (16) included the treatment of fringing field effects.

The above refferences include only basic papers. Some more are added in refs. (17-23). An important methodical development represents the introduction by Penner (24) of matrix methods for dealing with bending of beams having small spatial and angular extent, and small momentum spread. The Equations of particle trajectories become then linear in these three parameters and can be expressed in a matrix formalism.

Sector beta spectroscopes. One of the earliearst studies of beta spectrometers of prismatic type was made by Siday and collaborators (25-29). They considered the motion in median plane of a field produced by circular plane poles parallel to each other. A brief description of such a spectrometer will be given bellow.

Two uniform field sector spectrometers will be described below. One of them, made by Berlovitch (30) achieves higher order radial focusing by shaping the boundaries, while another made Paris (31) uses special entrance and exit angles.

Several double-focusing sector spectrometers were constructed by Sakai, Ikegami and others (32-35) in Tokyo. Ikegami also considered theoretically the case of curved boundaries. Kaminskii and Kaganskii (38) built a spectrometer similar to those in Tokyo. A double-focusing sector spectrometer with a deflection angle of 60° was made by Bhattacharya and Bhiday (39). De Jaeger, Douma, Bruinsma and de Vries (40) constructed a double-focusing spectrometer with a deflection angle of $(2/3) \pi \sqrt{2}$ radians, which they call "magic angle".

An important class of sector spectrometers are known as "orange". Besides complete multigap "orange", quite a few single gap sector spectrometers were made. They shall be considered separately in chapter 12.

Basic focusing properties will be derived in the following for the simple case of uniform field sector, with straight boundaries and neglecting the fringing field effects. It will be assumed in all cases that the central ray is entering and leaving the sector perpendicularly to its boundaries. More general case of oblique entrance is treated in references (1-2). The formulae are then much more complicated and do not convey so directly the dependance of focusing on the basic parameters. Moreover the perpendicular entrance is generally of more interest for electron beams.

Focusing in uniform field sectors

Barbers rule. In the early days of development of mass spectrometers, Barber (4) found a simple rule describing a geometrical property of sector magnets with uniform field, valid for the case when the beam enters and leaves the sector perpendicularly to pole-pieces. The focusing is considered in the plane perpendicular to the field and the concept used - point source and focus - are characteristic for first order treatment. The rule states that the center of the circle, described in the sector by central trajectory, is collinear with the source and the focus, independently of the sector angle.

We will reproduce a simple proof of Barbers rule given by Cartan (7). He starts with the limiting case of sector angle equal to 180°, corresponding to semi-circular spectrometer and shows that the focusing angle cannot be other but 180°. Suppose that the focusing angle is smaller than 180° and that the focus is in the point F' (Fig.1a). The centers of the orbits arriving to F' should lie on an arc C' having F' as center. Considering now the starting point S of the same orbits, it is seen that the centers of the circles should lie on the arc C described from S. Since the two arcs C and C' contain centers of the same circular orbits, they should coincide. This is possible in the first order approximation, only when the two arcs are tangent, corresponding to the focusing angle of 180°. Then the points S, O, and F are collinear.

Passing now to the sector angle smaller than 180°, we consider a central ray entering the magnet at 90° and a ray also emitted from S (Fig.1.b) at a small angle ζ with respect to the central ray. The treatment is again two-dimensional and the fringing field is neglected. The orbits inside the prism are circular, tangent to the rays at entering points A and A'. In the first approximation it can be taken that the side of the prism has a circular shape with the point source S as center. The centers of circular orbits starting tangentially at A and A' should be lying on an arc described from S. The same argument can be repeated for the exit side. The arc described from F should contain centers of circular orbits tangent to the straight lines BF and B'F. Since the circular orbits are the same, the two arcs should coincide, and this is possible in first approximation only if S, O and F are collinear.

Dispersion. We start from a central ray SABF produced by a particle having a momentum p and consider the trajectory of another particle having the momentum p - dp and entering the prism at the same point A. It will describe inside the prism an arc AB' having the radius $\rho+d\rho$ and crossing the SOF line at F' (Fig.2). The dispersion D can be defined as

$$D = FF'/(dp/p) \qquad\qquad (9.1)$$

It can be seen from Fig.2 that FF' is aproximately given by

$$FF' = FC/\cos \theta_2 \qquad\qquad (9.2)$$

which is obtained when the small angle ε is neglected compared to 90°. We shall first calculate FC which is equal to

$$FC = b + x \qquad\qquad (9.3)$$

where $b = BB'$.

In the triangle OO'B' the sides $\rho + b$, $\rho + d\rho$ and d are connected with the angle ($\phi - \varepsilon$) at O' by the relation

$$(\rho + b)^2 = (\rho + \rho)^2 + d\rho^2 - 2(\rho + d\rho)\, d\rho \cos(\phi - \varepsilon)$$

which by neglecting small magnitudes of second order becomes

$$b = d\rho \ (1 - \cos \Phi) \tag{9.4}$$

In the same triangle also holds the relation

$$\frac{d\rho}{\sin \varepsilon} = \frac{\rho + d\rho}{\sin (\pi - \phi)}$$

which, when neglecting small magnitudes of second order gives

$$\varepsilon = \frac{d\rho}{\rho} \ \sin \phi \tag{9.5}$$

It can be seen from Fig.2. That

$$x = \ell_2 \tan \varepsilon$$
$$\approx \ell_2 \varepsilon$$

so that using (5), x becomes

$$x = \frac{d\rho}{\rho} \ \ell_2 \sin \phi \tag{9.6}$$

Adding (4) and (6), FC becomes

$$FC = \frac{d\rho}{\rho} \ \left| (1 - \cos \phi) \rho + \ell_2 \sin \phi \right| \tag{9.7}$$

Since $d\rho/\rho = dp/p$, the above relation can be written as

$$FC = dp/p \ \left| (1-\cos \phi) \rho + \ell_2 \sin \phi \right| \tag{9.8}$$

The quantity inside brackets can be transformed to

$$(1-\cos \phi) \rho + \ell_2 \sin \phi = \rho \ (1-\cos \phi + \sin \phi \tan \theta_2)$$
$$= \rho \ (1-\cos (\phi + \theta_2)/\cos \theta_2)$$
$$= \rho \ (1+\cos \theta_1/\cos \theta_2) \tag{9.9}$$

Dispersion then becomes

$$D = \frac{\rho}{\cos \theta_2} \ (1+\cos \theta_1/\cos \theta_2) \tag{9.10}$$

The expressions (8) and (10) are often used to caracterise the dispersion. The dispersion can be expressed as function of linear dimensions by noting that

$$\cos \theta_2 = \frac{\rho}{y_f} \quad \cos \theta_1 = \frac{\rho}{y_s} \tag{9.11}$$

and eliminating θ_1 and θ_2 from (10), one obtains

$$D = (y_s + y_f) \frac{y_f}{y_s} \tag{9.12}$$

The dispersion increases with linear distance of the apparatus characterised by the source to the detector distance $(y_s + y_f)$ and for given linear dimensions increases with the ratio of y_f/y_s.

For comparison with other spectrometers one has to take FC as measure of dispersion, and using D_r rather than D, (10) gives

$$D_r = (1 + \cos\theta_1/\cos\theta_2)$$
$$= (1 + y_f/y_s) \tag{9.13}$$

In the case of a symmetric spectrometer (13) becomes

$$D_r = 2$$

which is also the dispersion of semi-circular spectrometer.

Image width. Two rays emitted at small angles $\pm\varepsilon$ with respect to central ray (Fig. 3) fall at a distance $FF´$ from the point F where the central ray reaches the SO-axis. The length $FF´$ represents the image width of a point source, for the two-dimensional case here considered. The image width in the direction perpendicular to central ray FF'' is connected with $FF´$ by the relation

$$FF´ = FF''/\cos\theta_2 \tag{9.14}$$

One can see from the Fig. 3 that FF'' is given by

$$FF'' = FN´ - BB´$$
$$= \ell_2 + \tan\eta\varepsilon - (OB´ - \rho) \tag{9.15}$$

From the properties of triangles $OB´O´$ and $OO´A´$, the following relations can be written

$$OB´ = \rho(\cos\varepsilon + \sin\varepsilon \cot\nu) \tag{9.16}$$
$$OA´ = \rho(\cos\zeta + \sin\zeta \cot\mu) \tag{9.17}$$

$$\frac{\sin\varepsilon}{\sin\nu} = \frac{\sin\zeta}{\sin\mu} \tag{9.18}$$

where

$$\zeta = \mu + \nu \tag{9.19}$$

With the help of (17) and (18) the relations (16) and (17) can be transformed to

$$OB' = \rho \left[\cos \quad - \cos(\zeta + \phi) + \frac{OA'}{\rho} \cos \phi \right] \tag{9.20}$$

$$OA' = \frac{\rho}{\sin \phi} \left[\sin(\zeta + \phi) + \sin \epsilon \right] \tag{9.21}$$

From the triangle SAA' one can write OA' as

$$OA' = \rho + AA'$$

$$= \rho + \ell_1 \tan\zeta \tag{9.22}$$

Since ζ is very small, one can develop $\cos \zeta$ and $\sin \zeta$, neglecting small magnitudes of third and higher orders. Passing to source and focus distances y_s and y_f, the relations (21) and (22) give then

$$\sin \epsilon = \zeta \frac{y_s}{y_f} + \zeta^2 \frac{\rho}{y_s y_f} (\ell_1 + \ell_2) \tag{9.23}$$

from which follows

$$\cos \epsilon = 1 - \frac{1}{2} \zeta^2 \frac{y_s^2}{y_f^2} \tag{9.24}$$

With help of (21-24), the relation (20) becomes

$$OB' = \rho \left[1. - \frac{\rho^2}{2} \frac{y_s^2}{y_f^2} + \frac{\rho^2}{2 y_s y_f} (\ell_1 \ell_2 - \rho^2) + \frac{\zeta}{\rho} \frac{y_s}{y_f} \ell_2 \right] \tag{9.25}$$

Since for small ϵ one can use (23) for $\tan\epsilon$, the relations (15) and (25) give

$$BB'' = \frac{\zeta^2}{2} \rho \left[\frac{y_s^2}{y_f^2} + \frac{y_f}{y_s} \right] \tag{9.26}$$

The relation (26) shows that the image width is proportional to the square of the aperture angle and increases with y_s / y_f ratio. For $\phi = 180^o$, (26) reduces to $\rho \zeta^2$, as required for semi-circular spectrometer.

Magnification. A line source, lying perpendicularly to central ray ℓ_1 (Fig. 4) and having a length b', produces a line image, having a length b''. The magnification is measured by ratio b''/b'

$$M = b''/b' \tag{9.27}$$

The image length can be written as

$$b'' = F'E - FE \tag{9.28}$$

From triangles $O'B'C$ and $BB'C$ one can deduce

$$d\Phi = \frac{b'}{\rho}\, \sin\Phi \tag{9.29}$$

which gives for $F'E$

$$F'E = \ell_2 \tan d\Phi \approx \ell_2 \cdot d\Phi$$

$$= \frac{\ell_2 \cdot b'}{\rho}\, \sin\Phi \tag{9.30}$$

From triangle $BB'C$ it follows that

$$FE = BB' = b'\cos\Phi \tag{9.31}$$

Putting (30) and (31) into (28), b'' becomes

$$b'' = b' \left| \frac{\ell_2}{\rho}\, \sin\Phi - \cos\Phi \right| \tag{9.32}$$

and magnification is given by

$$M = \frac{b''}{b'} = \left| \frac{\ell_2}{\rho}\, \sin\Phi - \cos\Phi \right| \tag{9.33}$$

Passing from angles to source and detector distances (33) becomes

$$M = y_f/y_s \tag{9.34}$$

which is of course equal to one, for the symmetric case.

By comparing (34) and (12) one can see that the dispersion and magnification are connected by the formula

$$D_r = 1 + M \tag{9.35}$$

Sectors with inhomogenous field

Generalised Barber's rule. Barber's rule can be generalised to the more interesting case of a sector with inhomogenous magnetic field. The rule will be derived making the following assumptions:

- The field inside the sector is cylindrically symmetric.

- The sector is wedge-shaped with straight boundaries, as shown on Fig. 5.

- The magnetic field is confined to the sector and fringing effects neglected.

- The beam enters and leaves perpendicularly to the sector sides.

The optic axes is defined by a ray SABD entering perpendicularly and following the radius r_o. Other trajectories can be characterised by $\eta = (r - r_o)/r_o$ and $\tau = z/r_o$. The motion of electrons inside the sector is governed by equations 4.45 and 4.46.

$$\eta'' + \omega_r \eta = 0 \tag{9.36}$$

$$\tau'' + \omega_z \tau = 0 \tag{9.37}$$

Consider a particle leaving the source at an angle ζ and moving in the median plane. The general solution of (35) is

$$\eta = C_1 \cos \omega_r \theta + C_2 \sin \omega_r \theta \tag{9.38}$$

The initial conditions, at the point A, are

$$\theta = 0 \quad \eta = \eta_1 \quad d\eta/d\theta = \eta_1' \tag{9.39}$$

so that (38) becomes

$$\eta = \eta_1 \cos \omega_r \theta + (\eta_1'/\omega_r) \sin \omega_r \theta \tag{9.40}$$

Since ℓ_2 depends on the derivative of η at B, we need $\eta = d\rho/d\theta$

$$\eta = -\eta_1 \omega_r \sin \omega_r \theta + \eta_1' \cos \omega_r \theta \tag{9.41}$$

At B, where $\theta = \Xi$, η and η become

$$\eta_2 = \eta_1 \cos \omega_r \Xi + (\eta'_1 / \omega_r) \sin \omega_r \Xi \tag{9.42}$$

$$\eta'_2 = \eta_1 \omega_r \sin \omega_r \Xi + \eta'_1 \cos \omega_r \Xi \tag{9.43}$$

From triangles SAA´ and FBB´ one finds that

$$\eta'_1 = r_o \eta_1 / \ell_1 \tag{9.44}$$

$$\eta'_2 = r_o \eta_2 / \ell_2 \tag{9.45}$$

Eliminating η_1, η_2, η'_1 and η'_2 from (42) and (43), te relation

$$\omega_r \sin \omega_r \Xi - (r_o / \ell_1) \cos \omega_r \Xi = (r_o / \ell_2) \cos \omega_r \Xi + (r_o^2 / \ell_1 \ell_2) \sin \omega_r \Xi \tag{9.46}$$

is obtained, which can be written

$$\frac{(\omega_r \cdot \ell_1 / r_o) + (\omega_r \ell_2 / r_o)}{(\omega_r \ell_1 \ell_2) / r_o^2 - 1} = \tan \omega_r \tag{9.47}$$

This can be transformed to

$$\text{arc tan}(\omega_r \ell_1 / r_o) + \text{arc tan}(\omega_r \ell_2 / r_o) + \omega_r \Xi = \pi \tag{9.48}$$

The equation for the axial focusing is quite similar

$$\text{arc tan}(\omega_z \ell_1 / r_o) + \text{arc tan}(\omega_r \ell_2 / r_o) + \omega_z \Xi = \pi \tag{9.49}$$

These equations represent a generalised form of Barber´s rule.
This can be seen by writting

$$\text{arc tan}(\omega_r \ell_1 / r_o) = \theta_1 \tag{9.50}$$

$$\text{arc tan}(\omega_r \ell_2 / r_o) = \theta_2 \tag{9.51}$$

$$\omega_r \Xi = \phi \tag{9.52}$$

Then (48) becomes

$$\theta_1 + \theta_2 + \phi = \pi \tag{9.53}$$

This relation is schematically shown in Fig. (5b).

Simple special cases.　　a)　Uniform field. - In this case

$$\alpha_1 = 0$$

$$\omega_r = \sqrt{1 + \alpha_1} = 1$$

$$\omega_r = \sqrt{-\alpha_1} = 0$$

and (48) becomes

$$\text{arc tan } (\ell_1/r_o) + \text{arc tan } (\ell_2/r_o) = \pi - \Xi \qquad (9.54)$$

In the case of symmetrical arrangment, $\ell_1 = \ell_2 = \ell$　　and　(49)

gives

$$\ell/r_o = \tan (\pi/2 - \Xi/2) \qquad (9.55)$$

This relation also follows from simple properties of triangle SAD (Fig. 5b).

b)　$\pi\sqrt{2}$　spectrometer. - In the case of Siegbahn–Svartholm double focusing spectrometer, the parameters are

$$\alpha_1 = -0.5$$

$$\omega_r = \omega_z = 1/\sqrt{2}$$

so that (48) becomes

$$\text{arc tan } (\ell_1/\sqrt{2}\, r_o) + \text{arc tan } (\ell_2/\sqrt{2}\, r_o) = \pi - \Xi/\sqrt{2} \qquad (9.56)$$

In the symmetrical case (20) reduces to

$$\ell = \sqrt{2}\, r_o \tan (\Xi/2\sqrt{2}) \qquad (9.57)$$

Magnification.　Generalised Barber´s rule indicates that the radial magnification M_r should be

$$M_r = -y_f'/y_s'$$

$$= \left[\frac{r_o^2 + (1 + \alpha_1)\, \ell_2^2}{r_o^2 + (1 + \alpha_1)\, \ell_1^2} \right]^{1/2} \qquad (9.58)$$

This relation can also be simply derived from the lens properties of the prism (see Judd (15)).

Dispersion. In the case of a uniform field, the dispersion D_r and (radial) magnification M_r are connected by the relation (35)

$$D_r = 1 + M_r \qquad\qquad (9.59)$$

giving $D_r = 2$ for semi-circular spectrometer. Since $D_r = 2$ follow from the general formula (4.70) giving

$$D_r = 2/(1 - \alpha_1) \qquad\qquad (9.60)$$

and (58) should reduce to (59) for the symmetric case, one can expect that a general formula for dispersion should read

$$D_r = \frac{1 + M_r}{1 - \alpha_1} \qquad\qquad (9.61)$$

This can of course be proved by more rigorous treatment.

As could be expected the dispersion increases with magnification.

Primatic beta-ray spectrometers

Circular pole prism spectrograph. Siday and collaborators (25-29) studied and constructed a prismatic beta-ray spectrometer having circular poles separated by a distance equal to the radius of either pole. The motion is confined to the neighborhood of the median plane, the axial opening being only 2°. Theoretical considerations of Siday (25) and Jennings (26), and experimental work with an electronic model apparatus made by Ehrenberg and Jennings (27) showed that second order radial aberration term can be eliminated and third order reduced for some entrance and deviation angles.

The magnet (28) has the radius and the separation of pole pieces equal to 10 cm. Fig.6 shows the geometry. The source to detector distance is about 40 cm. A photographic plate accepts a spectrum range of 20% in momentum. With a source of 0.1×20 mm^2, a line width of 0.04% was obtained for a radial opening of 20°.

Curved boundary, uniform field spectrometer. Berlovich described two sector spectrometers (30) constructed for coincidence work. Both
have uniform field and shaped pole-piece boundaries, the second-one representing an ameliorated version of the first-one. The geometry of the second sector is shown in Fig.7. On the entrance side the magnet boundary has a circular shape to aliminate the effect of cylindrical lens formed by fringing flux.
The exit boundary is so shaped as to bring the electrons in the median plane
to a focus, neglecting the fringing field. The accepted beam is wide and flat,
with a solid angle of 1%. The corresponding resolution is about 2%.

Oblique entrance, uniform field spectrograph. Paris (31) constructed a spectrograph with uniform field and sector angle equal to 180°. He
based the design on calculations made by Kerwin (8), showing that for an angle
of $54^{o}44'$ between the central ray and the edges of the pole-pieces, a second
order focusing is achieved in the median plane. Moreover, a straight focal line
is obtained in the field-free region (Fig.8).

In this type of spectrograph the transmission at low resolutions
is larger than in the semi-circular type, but at high resolutions, that is   not
the case any more due to the disturbing effects of the fringing field.

Tokyo sector double-focusing spectrometers. Sakai and collaborators (32-34) started the construction of sector double focusing spectrometers in 1956, and several such instruments (INS-I to INS-IV) were developed in Institute for Nuclear Studies. The spectrometer became later commercially available from Hitachi Ltd., and we shall summarise below the extensive description of the latest model, given by Yamamoto, Takumi and Ikegami (35).

The characteristic parameters of the spectrometer are the following:

$$B = B_{o}(1 - 0.5\eta + \frac{1}{4}\eta^{2} - \frac{1}{8}\eta^{3})$$

$$r_{o} = 34 \text{ cm}$$

$$2z_{o} = 24 \text{ cm}$$

$$\Xi = 194^{o}$$

$$\ell_{1} = 44.7$$

$$\ell_2 = 0$$
$$M_r = 0.733$$
$$D_r = 3.47$$

The geometry is shown in Fig.9. The source is outside of the magnet, to facilitate the coincidence work. The foucus is inside the field, to avoid the distorsions due to fringing flux and leave enough space for second instrument.

The magnet has the iron core surrounded by six coils, and an outer yoke, which serves also as vacuum chamber wall, with another six coils, having equal number of amper-turns as inner-ones. Iron shims are placed between the coils inside and outside, facing each other and forming magnetic equipotential surfaces. The best current distribution in the coils was found empirically.

The relatively large gap width of $2z_o = 24$ cm compared to source distance $\ell_2 = 44,7$ cm, makes the fringing field problem more complex, than it is for small gaps. The fringing field effects were handled semi-empirically, in two steps:

- First, the approximate theoretical fringing field distributions were integrated, to obtain the effective field boundary defined as a discontinuous boundary which produces equivalent deflection of electrons. Fig.10 shows four theoretical fringing field distributions and corresponding trajectories in the median plane. They yield approximate values for focal distance, necessary for the design.

- Once the magnet is constructed and the field is measured, the source position was found by calculating the trajectories and by ray-tracing. Fig.11 shows the distribution of the fringing field along the reference trajectory compared to the theoretical field (No.4).

The authors give the following performance data:

Source, mm^2	R, %	T, %	L, cm^2
1 x 10	0,08	0,01	10^{-5}
1 x 15	0,2	0,2	3.10^{-4}

A convenient feature of the spectrometer is that a focal line exists (Fig. 12) covering 15% of the momentum range. The resolution differs over the focal line for about 15%.

Kaminskii and Kaganskii (38) constructed an iron-core double-focusing spectrometer similar to those developed by Sakai and collaborators.

The main characteristics of the spectrometer are the following:

field coefficients $\alpha_1 = -0.5$ $\alpha_2 = 3/8$

$r_o = 25$ cm

$2z_o = 14$ cm

$\ell_2 = 29$ cm

effective sector angle 200°

The source is placed inside the magnet (Fig. 13) so that the beam has to pass the fringing field only once. The final adjustment of the field form inside the magnet is made with help of two pairs of shims, one inside and another outside of the optic circle.

The performances of the spectrometer are given in the table below.

Table I

Source size mm^2	Slit width mm	Solid angle %	Resolution %
1 x 20	12	0,9	1,3
1 x 20	3	0,3	0,36

Magic-angle electron spectrometer at Amsterdam. De Jager, Douma, Bruinsma and de Vries (40) have described an electron spectrometer used for electron scattering experiments with the 85 MeV linear electron accelerator. The disign is based on a study made by Penner (41), who proposed the focusing angle $\Xi = (2/3\ \pi\sqrt{2} = 169.7^o$, called by authors "magic angle". Second order radial focusing is independent of ℓ_1, and for a given ℓ_1, the magnet to image distance ℓ_2 is twice as large as for the $\Xi = 180^o$ double-focusing spectrometer, leaving more space for the counting apparatus.

The characteristic parameters of the spectrometer are:

$$\alpha_1 = -0.5$$

$$\alpha_2 = 0.25$$

$$r_o = 65 \text{ cm}$$

$$2z_o = 6 \text{ cm}$$

$$\Xi = 169,7^\circ$$

$$\ell_1 = 75 \text{ cm}$$

$$\ell_2 = 36 \text{ cm}$$

$$(B_o)_{max} = 12.5 \text{ kG}$$

As can be seen on Fig.14, in order to obtain the effective bending angle of $169,7^\circ$, the angle of the sector itself is $163^\circ 20'$.

The field measurements gave for field coefficients $\alpha_1 = -0.495 \pm 0.002$ and $\alpha_2 = 0.246 \pm 0.006$. The field coefficients were found to be independent of the field strength.

The results of fringing field measurements are shown in Fig.15. It can be seen that the fringing field is slightly shorter for higher fields, so that the effective bending angle is reduced from 170.4° at 1.8 kG, to 169.8° at 11 kG.

The resolution of the spectrometer, is 0.1% at the solid angle of 0.05%.

References

1. H. Ewald, H. Hintenberger, Methoden and Anwendungen der Massenspektro-skopie, Chemie, Weinheim, 1953.

2. K.T. Bainbridge, In "Experimental Nuclear Physics" (E. Segre, ed.), Vol. I, p. 559, Willey, New York, 1953.

3. H.A. Enge, p. 203, and R. Casting, J.F. Hennequin, L. Henry, and G. Slodzian, p. 265, in "Focusing of charched particles", A. Septier, ed., Vol. II, Academic Press, New York, 1967.

4. N.F. Barber, Proc. Leeds Phil. Soc. 2, 427 (1933).

5. W.E. Stephens, Phys. Rev. 45, 513 (1934).

6. R. Herzog, Z. Phys. 89, 447 (1934).

7. L. Cartan, J. de Phys. 8, 453 (1937).

8. L. Kerwin, Rev. Sci. Instr. 20, 36 (1949), and 20, 381 (1949).

9. D.F. Dempsey, Rev. Sci. Instr. 26, 1141, 1955.

10. H. Hintenberger, Z. Naturforsch, 3a, 125 and 669 (1948); 6a, 275 (1951); Rev. Sci. Instr. 20, 748 (1949).

11. W.G. Gross, Rev. Sci. Instr. 22, 717, 1951.

12. M. Camac, Rev. Sci. Instr. 22, 197, 1951.

13. N. Svartholm, Ark. f. Fysik, 2, 115 (1950).

14. E.S. Rosenblum, Rev. Sci. Instr. 21, 586 (1950).

15. D.L. Judd, Rev. Sci. Instr. 21, 213 (1950).

16. D.L. Judd and S.A. Bludman, Nucl. Instr. Meth. 1, 46 (1957).

17. M. Spiegel, J. Phys. Rad. 10, 204 (1949).

18. L. Kerwin and G. Geoffrion, Rev. Sci. Instr. 20, 381 (1949).

19. L. Musemuci, Nuovo Cim. 7, 351 (1950).

20. M. Cotte, Ann. Phys. 10, 333 (1938).

21. C. Berry, Rev. Sci. Instr. 27, 849 (1956).

22. R.M. Sternheimer, Rev. Sci. Instr. 24, 573 (1953).

23. Ju. Basargin, Zhurn. Tekn. Fiz. 37, 1360 (1967).

24. S. Penner, Rev. Sci. Instr. 32, 150 (1961).

25. R.E. Siday, Proc. Phys. Soc. 59, 905 (1947).

26. J.C.E. Jennings, Proc. Phys. Soc. B, 65, 256 (1952).

27. W. Ehrenberg and J.C.E. Jennings, Proc. Phys. Soc. B, 65 265 (1952).

28. R.E.Siday and P.A.Silverston, Proc. Phys.Soc. A 65, 328 (1952).

29. R.E.Siday, Physica 18, 1063 (1952).

30. Je.E.Berlovitch, Izvestia AN SSSR, Ser. fiz. 18, 589 (1954).

31. P.Paris, Journ. Rech. C.N.R.S. No. 54, 1961.

32. M.Sakai, H.Ikegami and T.Yamazaki, Nucl.Instr.Meth. 9, 154 (1960), and 25, 328 (1964).

33. M.Sakai, Nucl.Instr.Meth. 8, 61 (1960).

34. M.Sakai and H.Ikegami, J.Phys. Soc. Japan 13, 1076 (1958).

35. H.Yamamoto, K.Takumi and H.Ikegami, Nucl.Instr.Meth. 65, 253 (1968).

36. H.Ikegami, Rev.Sci.Instr. 29, 943 (1958).

38. D.L.Kaminskii and M.G.Kaganskii, Prib.Tekn.Eksp. (1959), No.1, 32.

39. P.K.Bhattacharya and M.R.Bhiday, Ind. J.Pure App. Phys. 5, 386 (1967).

40. C.W. de Jager, F.Th. Douma, P.J.T.Bruinsma and C. de Vries, Nucl. Instr. Meth. 74, 13 (1969).

41. S.Penner, Rev.Sci. Instr. 32, 150 (1961).

Text to Figures. Ch.9

Fig. 9.1.   Derivation of Barbers rule (Cartan). (a) Semi-circular case.
(b) General case.

Fig. 9.2.   Derivation of dispersion for simple case of ortogonal entrance.

Fig. 9.3.   Derivation of image width.

Fig. 9.4.   Derivation of magnification.

Fig. 9.5.   Derivation of generalised Barbers rule. (a) notation, (b) Schematic
presentation of focusing relations.

Fig. 9.6.   The geometry of Siday´s prismatic spectrograph.

Fig. 9.7.   The geometry of curved boundary, uniform field spectromewr
(Berlovich).

Fig. 9.8.   The geometry of oblique entrance, uniform field spectrometer
(Paris).

Fig. 9.9.   The geometry of INS sector double-focusing spectrometer.

Fig.9.10.   (a) Four theoretical fringing field distributions, (b) corresponding
trajectories in median plane.

Fig.9.11.   The fringing field along the reference trajectory, compared to the
theoretical field.

Fig.9.12.   The focal line of INS spectrometer.

Fig.9.13.   The geometry of sector double - focusing spectrometer with the
source in side the field. (Kaminskii, Kaganskii).

Fig.9.14.   The geometry of "magic angle" spectrometer.

Fig.9.15.   The fringing field of "magic angle" spectrometer.

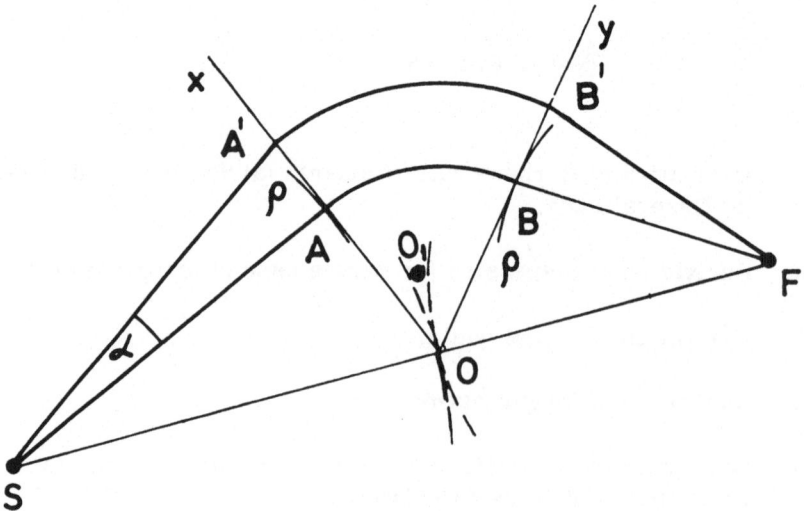

9.1. a

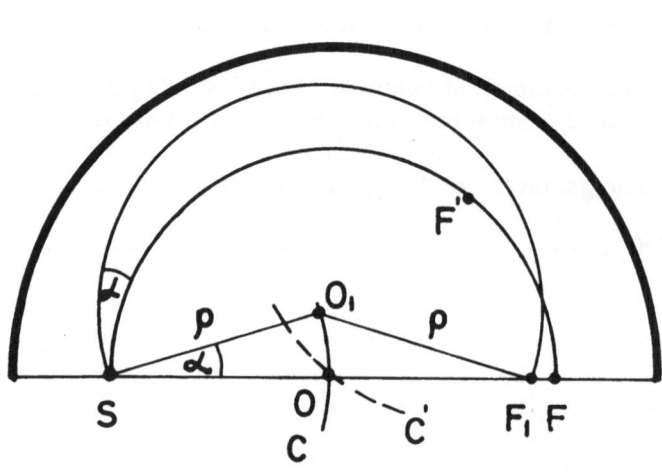

9.1. b

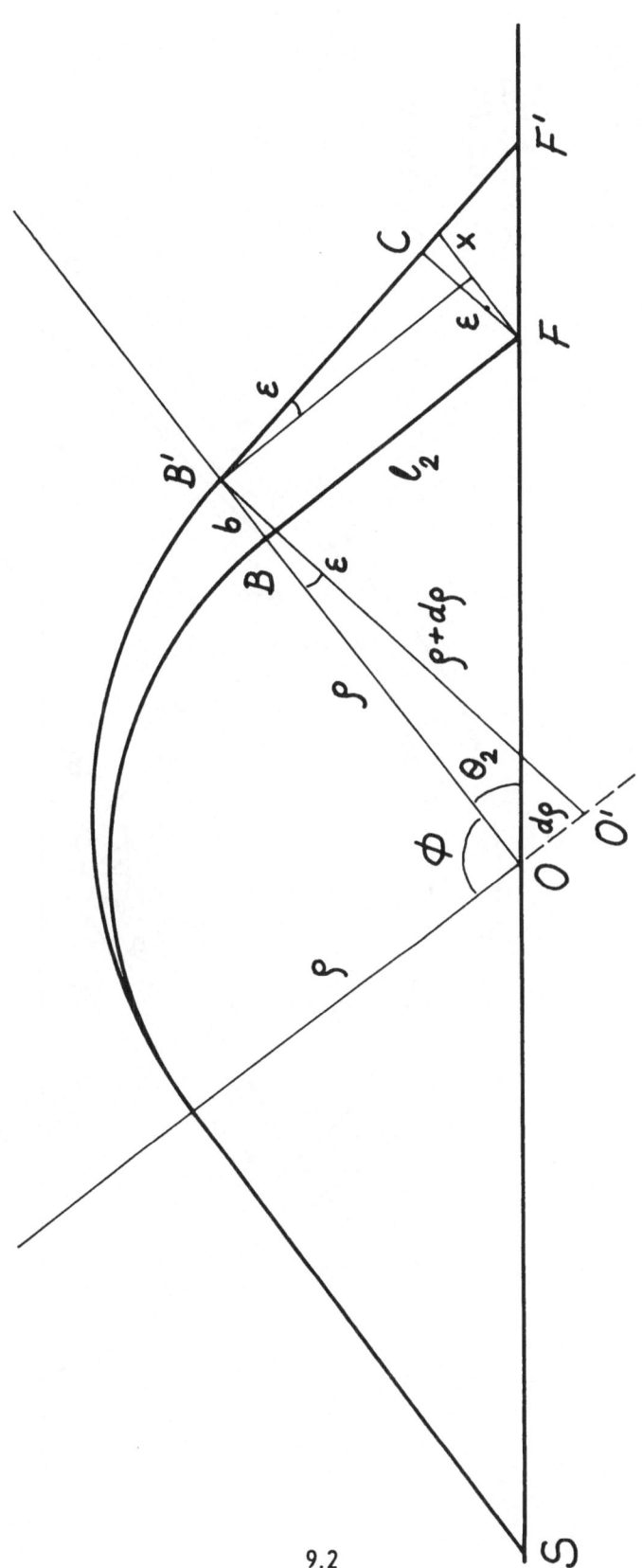

9.2

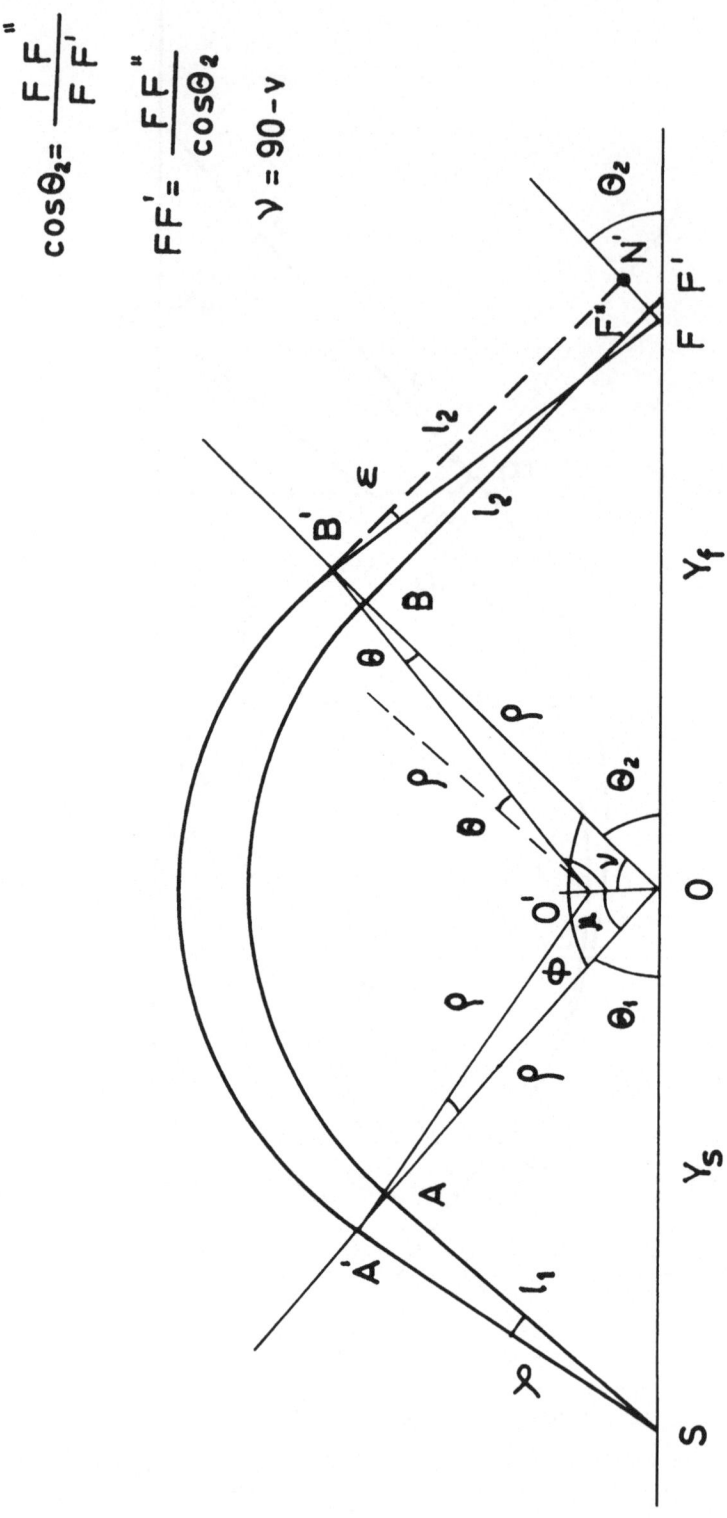

$$\cos\Theta_2 = \frac{FF''}{FF'}$$

$$FF' = \frac{FF''}{\cos\Theta_2}$$

$$\gamma = 90 - \nu$$

9.3

149

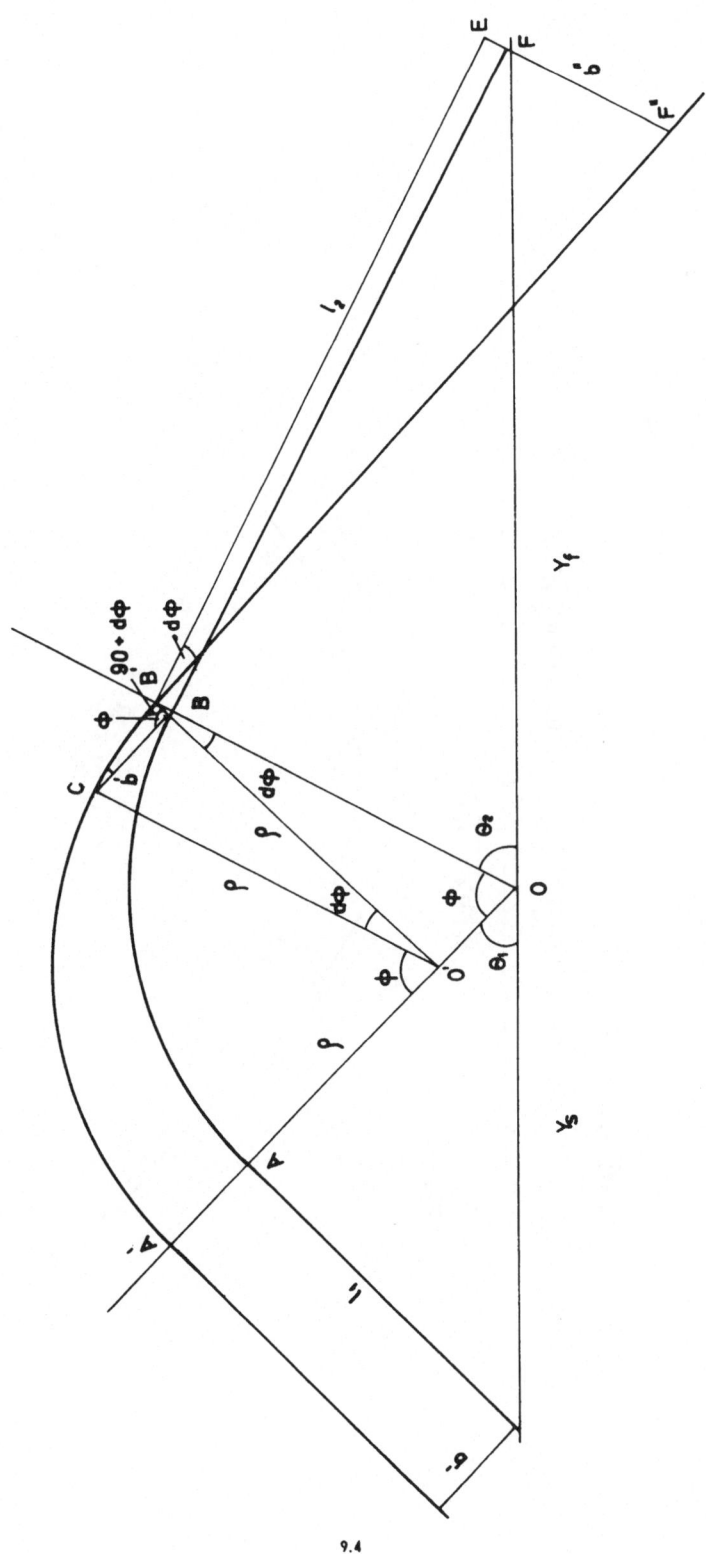

9.4

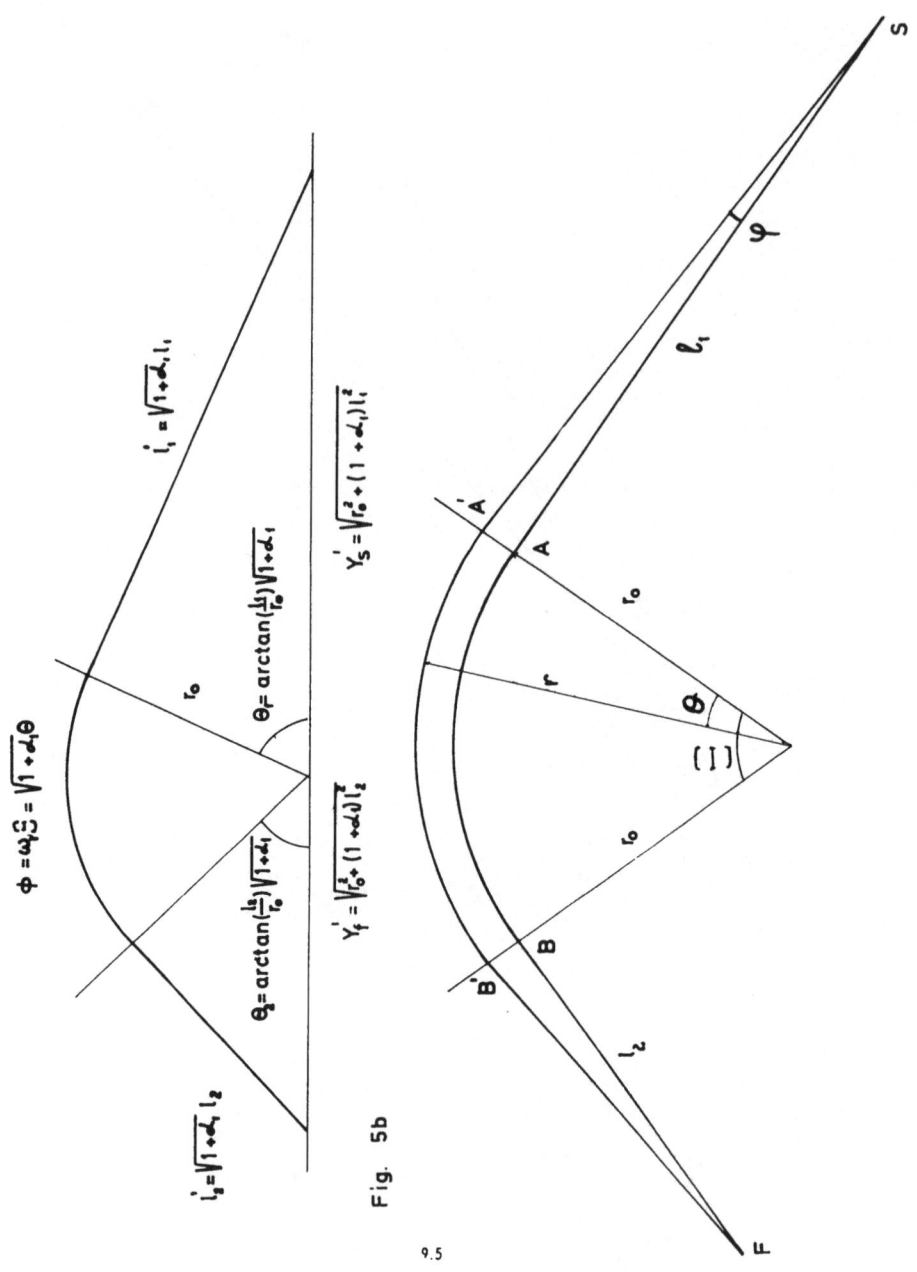

Fig. 5b

9.5

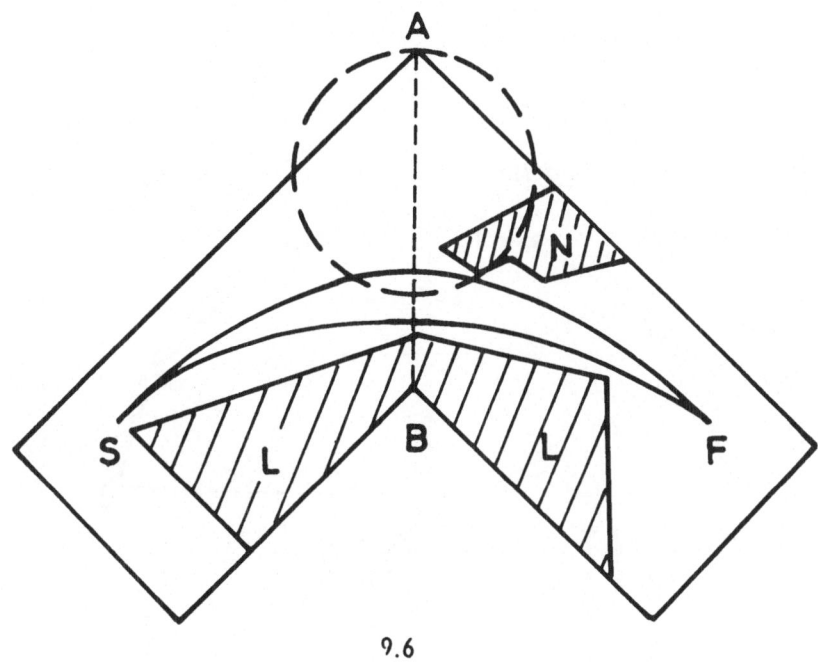

9.6

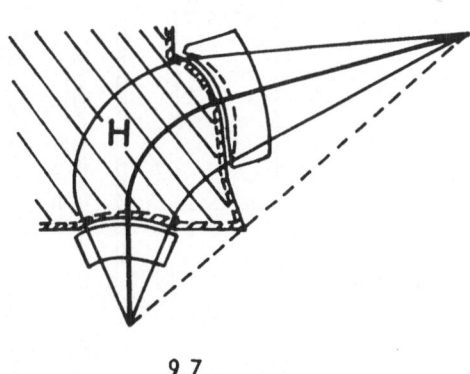

9.7

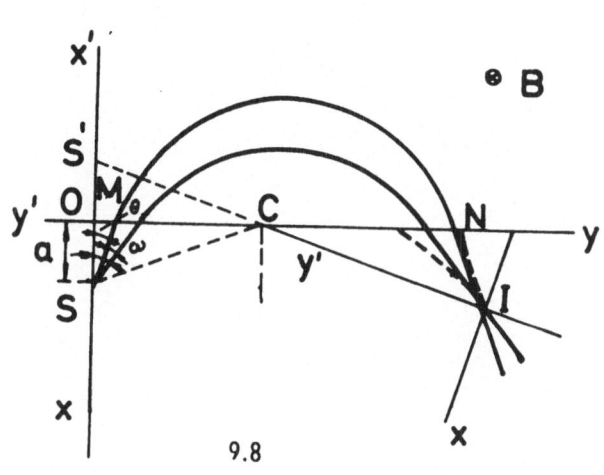

9.8

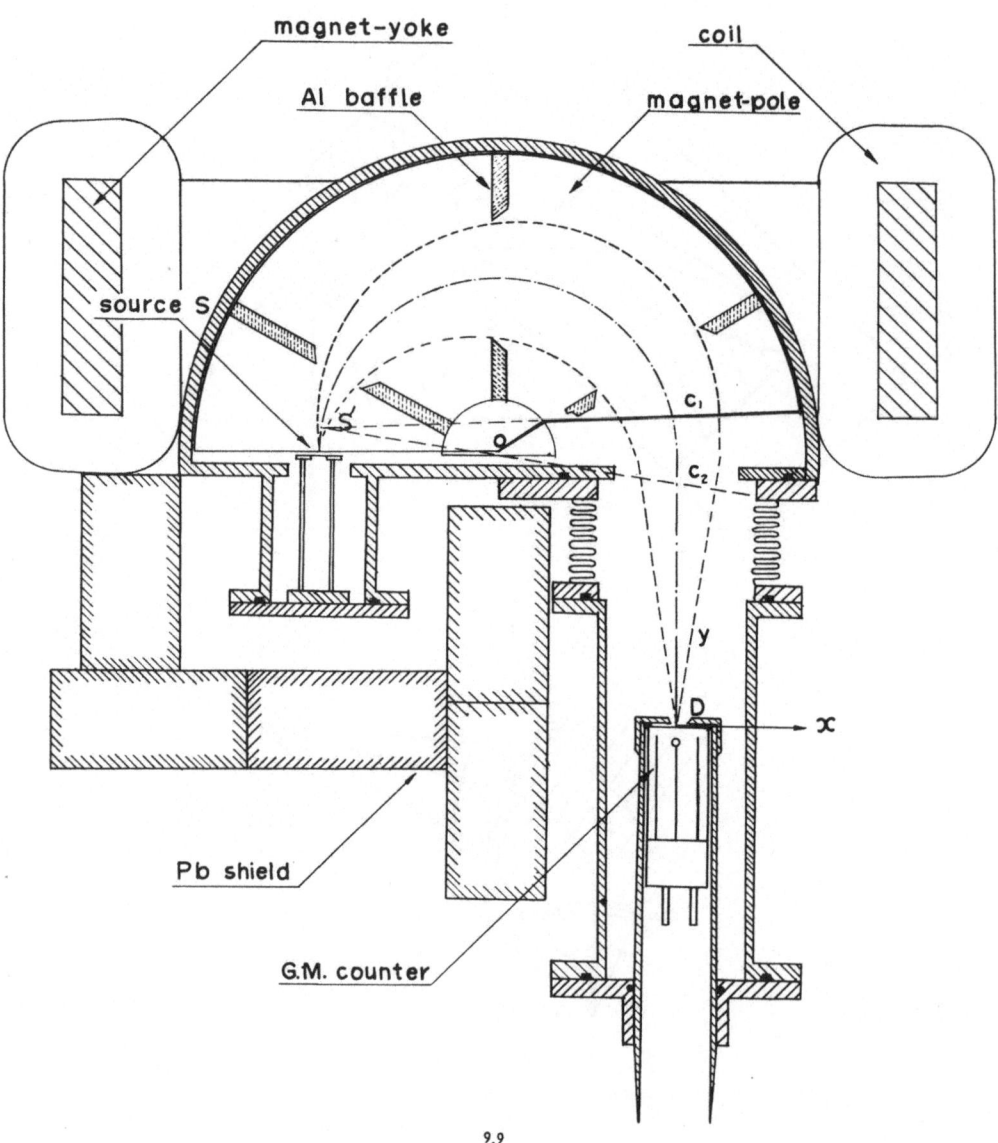

9.9

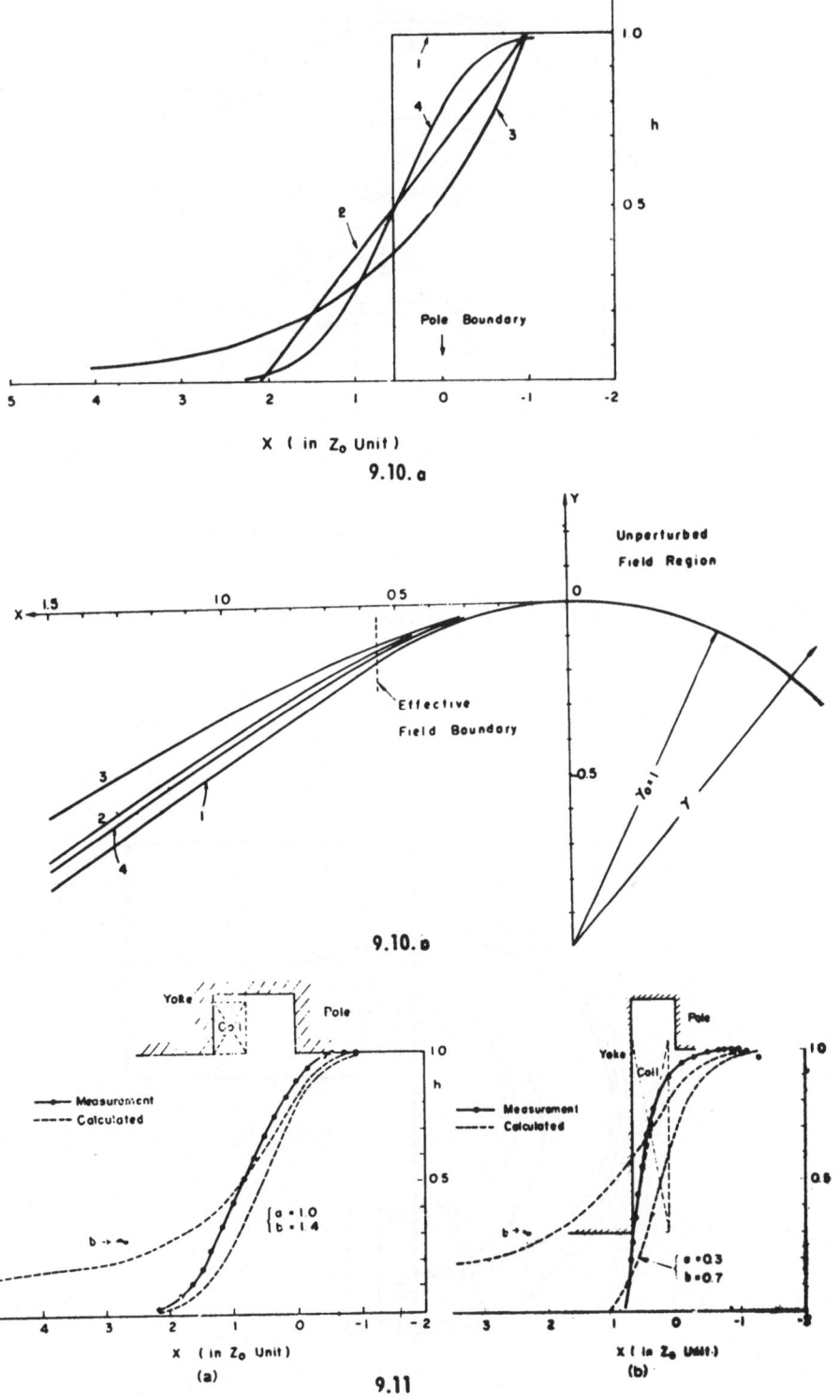

9.10. a

9.10. b

9.11

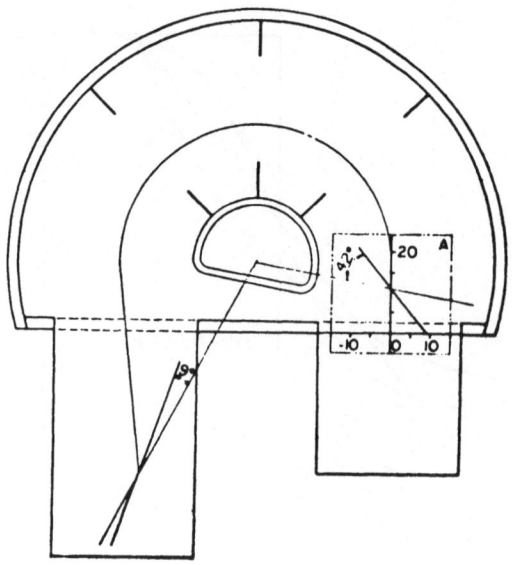

9.12

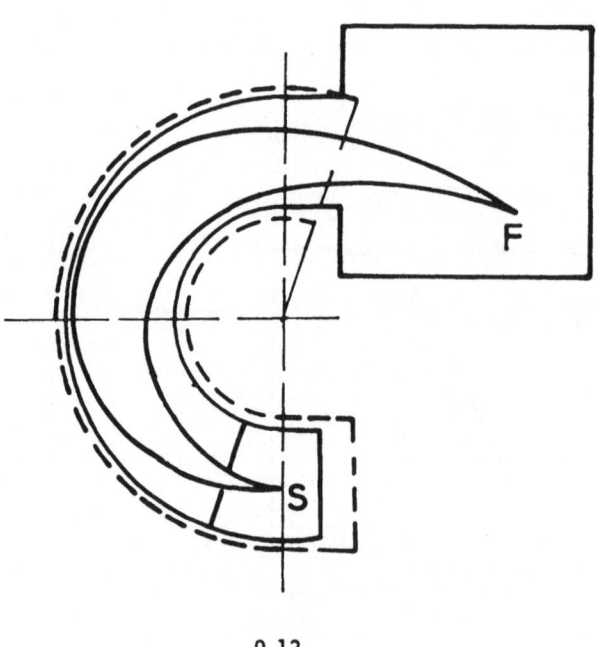

9.13

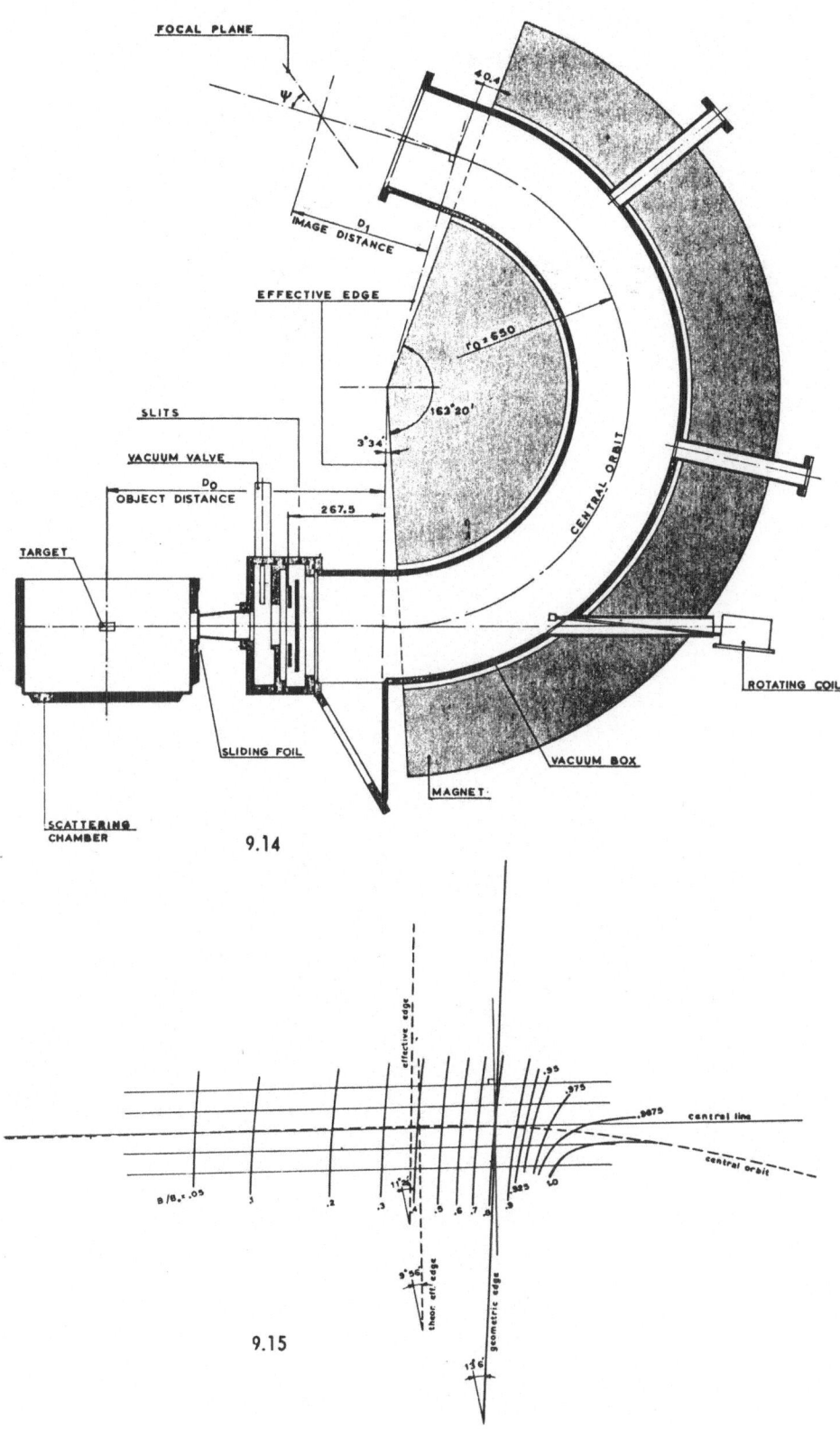

9.14

9.15

10. CORRECTORS AND NON-SYMMETRIC FIELDS

The succession of flat spectrometers, from semi-circular to $\pi/2 \sqrt{13}$, shows how performances are improved with more sophisticated field shapes and higher precision in magnet design, which essentially offers more degrees of freedom to handle the aberrations. So far only single magnets, with cylindrical symmetry were considered.

Another way to improve the performance is to give up the cylindrical symmetry and thus obtain new degrees of freedom to dispose of. This can be achived either by adding a new electron-optical device, magnetic or electro-static, to a cylindrically symmetric field or by using a single magnet with non-symmetrical field. The additional devices which we shall call correctors have often been conceived and designed after the spectrometer has been in use.

Improvements can be classified into three categories:

1. First order magnitudes improvements. The width of the source contributes to the line-width a term of first order. Bergkvist developed an electrostatic device (1) which allows to use large multi-strip sources without loss of resolution.

2. Higher order terms elimination. The contribution of higher order aberration terms can be eliminated either by correctors or in a very general way, by azimuthally varying fields (AVF). Appropriate correctors were developed by Walen (2) for a semi-circular spectrograph and Bergkvist (1) for a $\pi\sqrt{2}$ spectrometer. Only initial theoretical work was done so far on the AVF spectrometer. First proposal was made by Sessler (3) and developed by Sessler and Bergkvist (4), Schmutzler and Daniel (5), and by Tarantin (6).

3. Addition of a new focusing dimension. Before Siegbahn and Swartholm proposed $\pi\sqrt{2}$ spectrometer, another spectrometer having axial focusing also, was conceived and development started in Leningrad (7-10). It was called ketron and in a way represents first AVF spectrometer.

We shall now consider briefly all the above mentioned developments.

1. <u>Multistrip source</u>. Principle of Bergkvist's multistrip source is illustrated in Fig.1, which shows the projections of two orbits on median

plane. An electron emitted perpendicularly to the source at the point S_o which is on the optic circle, follows this circle, if the field has appropriate intensity. Another electron of the same momentum emitted perpendicularly from the point S_1 will cross optic circle and hit ϕ_o-plane at a point S_2. Since the magnification of $\pi\sqrt{2}$ spectrometer is -1, the distance DS_2 is equal to $-S_oS_1$, to the first order. The distance S_2D would contribute to the base line width the form $S_2D/4r_o$. This term could be eliminated if it were possible to arrange that electron emitted from S_1 with the same initial conditions, arrive to D. The deflection should be somewhat smaller, and that would happen if electron momentum were somewhat higher. That additional momentum is in Bergkvist´s device, communicated to electron by an electric field established between the source and a grid placed in front of it. The momentum of the electron increases from p_o to $p_o(1 + \Delta p)$. Since the fractional increase of the momentum should have the same magnitude as source width contribution to base line width $S_2D/4t_o$, it follows that

$$\Delta p = \frac{S_2D}{4r_o} = \tau_o/4$$

The accelarating potential V producing Δp is linearly proportional to the relative radial distance

$$V(\tau_o) = c_s$$

The constant c_s depends on the initial electron momentum p_o. All this simplified treatment was carried to lowest order and a more refined picture would require higher order corrections.

In principle potential should vary continuously allong the source, but in practice the variation is discrete, with source cut in a number of narow strips, each being at a different potential. Bergkvist used sources with 30 strips. Optimum focusing conditions are obtained, when the plane of the source is inclined 43° with respect to radius. The potential at 1 MeV amounts to 8 kV(Fig.2).

The best resolutions so far achieved with multi-strip source were about 0,04%. Inherent aberrations and imperfections of source geometry make if very hard to approach resolutions of 0,01%. The appropriate line widths to work with multistrip sources appear at present to be those above 0,05%.

The multiple-strip source is not limited to $\pi\sqrt{2}$ spectrometers but can be mounted in many other types of spectrometers. It has been construc- ted by Jahn for ($\pi/2$) $\sqrt{13}$ spectrometer (7) and by Peregud and collaborators (8) for optical analogy spectrometer.

2. Corrector for semi-circular spectrograph. One of the simplest correctors was designed by Walen (2), for a semi-circular alpha-ray spectro- graph. The purpose of the corrector was to eliminate the contribution of the square of radial opening angle ψ_r^2 to the line width, over an extended focal plane. No attempt were made to obtain axial focusing.

The design of the corrector was based on the property of semi- -circular spectrometers that the central ray hit the focal plane further away, than the two extreme rays. By increasing the field over a part of central ray trajectory, it would be more deflected and brought to the focus of extreme rays. The corrector is placed half-way between the source and the photographic plate. It consists of two iron plates having the shapes so calculated that the field becomes gradually stronger as the central part of the beam is approached (Fig. 3).

3. Correctors for $\pi\sqrt{2}$ spectrometers. The elimination of H_o^2 term in high aperture or T_o^2 term in wide aperture spectrometer line-widths can be obtained either with electrostatic or with magnetic correctors. Various possibilities are collected in Table 10.I.

Bergkvist was first to develop a corrector (1) for a $\pi\sqrt{2}$ spectro- meter. He has chosen the electrostatic type for a corrector in an iron-core spectrometer. The principle of electrostatic corrector is illustratede in Fig.4. It consists of two pairs of curved grids. The outer grids are at the potential of the spectrometer while the inner ones are at a negative potential. When passing through the first pair of grids, the particle is decelerated and in the second pair it is again accelerated for about the same amount. In both pairs there is a deflecting component of the electric field which increases with τ . The direc- tion of this deflecting component is such that the radius r of particle trajecto- ries is increased by an amount (rH^2) necessary to bring them into point focus. on the optic circle (in point-source approximation). The correcting potential V_c

is −3,2 kV for K-line of 412 keV transition in ^{198}Hg. The instalation of electrostatic corrector improved the resolution-transmission characteristics by a factor of four, which together with multistrip source amounts to a luminosity improvement by two orders of magnitude.

Lee-Whiting (9) calculated a cylindrical electrostatic corrector for iron-free $\pi\sqrt{2}$ spectrometer, which should improve the transmission by a factor of two. This figure could be increased with a non-cylindrical corrector which would work over the whole of the available aperture.

The advantage of electro-static correctors is that they can be developed semi-empirically. It is relatively simple to shape the wires and determine the right potentials. On the other hand, they represent separate electron-optical devices which have to be matched to magnets and need separate control and power supplies.

In principle it would perhaps be simpler to operate a spectrometer with a magnetic corrector, because correcting coils could be connected in series with spectrometer coils, while in the case of iron, additional pieces represent an integral part of the pole-pieces. The spectrometer then remains a single magnet device.

Shibata, Taya and Yoshizawa (10) developed a coil corrector for an iron-core $\pi\sqrt{2}$ spectrometer. As shown in Fig. 5. The coil is placed half-way between the source and detector. The field in the coil is opposite to the main field and the width of the coil increases with $|\tau|$ so that the deflection of extreme trajectories is decreased.

Simplified calculations of the corrector coil shape were made, valid for median plane and it was found that the approximate shape was given by two parabolae.

The corrector itself is shown in Fig. 6. All turns are identical, although one would expect that they should have an axial dependance. Author have found that this approximation does not introduce any defocusing effect.

The coil, being in the beam, decreases transmision by about 10%.

The advantage of the coil corrector is the possibility of series connection with main coils, but the presence of iron introduces a problem of non--linearity.

Authors have tested the relation between the optimum corrector current and the momentum of electrons. As could be expected the relation was not linear. They have nevertheless connected the currents in series because the optimum of the resolution - corrector current function is not sharp.

The effect of correction coil was illustrated by the conversion line of ^{137}Cs, the width of which was improved by a factor of three.

The design of iron-correctors for iron-core spectrometers, in principle similar to those made by Walen for semi-circular spectrograph, should not present much difficulties.

Sphalek described (11) a general approach to calculation of coil correctors for iron-free flat spectrometers. A practical problem is the return circuit. Shpalek assumed an infinite solenoid. A toroidal coil is also possible. It has to be large anyway, but the field is fortunately very small.

4. Ketron. Korsunskii, Kelman and Petrov (12) proposed a simple two-dimensional field to correct second order aberrations in a semi-circular spectrometer. It is a combination of homogenous field and field decreasing in one dimension. The source and the detector are in the homogenous field, while orbits pass partially through the inhomogenous field. We shall give simple theory of ketron following closely the book of Kelman and Yavor (13).

The theoretical treatment is confined to median plane, where x-axis passes through source and detector, the center of coordinate system being midway between them (Fig. 7), and aberration corrections are obtained by introducing a field gradient along positive y-axis.

Two orbits are shown in Fig. 7(a) one for electron emitted in the direction of y-axis and another making an angle $\Delta \zeta$ with respect to the first. In a semi-circular spectrometer, an electron emitted at an angle to y-axis return to x-axis at a distance smaller than $2\rho_0$. It can be made to return at a distance of $2\rho_0$, by having the field properly reduced along a part of its orbit. Suppose that the field is reduced at a distance $y = \rho_0$ from x-axis, and has a value B_1 from $y = \rho_0$ to $y = \rho_0 + \Delta y_1$. The inner orbit is completely in the homogenous field while outer enters from B_0 to B_1 at the point R. B_1 has to be chosen so that at point B the tangent to the orbit is parallel to the x-axis, which means that the center of the arc 0_1 is on the y-axis. The radius of the arc which electron describes in the region where the field is B_1, can be denoted by $\rho_1 = \rho_0 + \Delta\rho_1$. It can be seen from Fig. 7 that

$$\overline{SO} = \rho_0 = \rho_0 \cos \Delta\zeta + \Delta\rho_1 \cos \alpha_{11} \tag{10.1}$$

The angle α_{11} can be eliminated by noting that

$$\overline{OA} = \rho_0 = \rho_0 \sin \Delta\zeta + \rho_0 \sin \alpha_{11} \tag{10.2}$$

which gives

$$\sin \alpha_{11} = 1 - \sin \Delta\zeta \tag{10.3}$$

and introducing (3) into (1), one obtains

$$\Delta\rho_1 = \rho_0 \frac{1 - \cos \Delta\zeta}{\sqrt{2 \sin\Delta\zeta - \sin^2 \Delta\zeta}} \tag{10.4}$$

and

$$\rho_1 = \rho_0 \left(1 + \frac{1 - \cos \Delta\zeta}{\sqrt{2 \sin \Delta\zeta - \sin^2 \Delta\zeta}}\right) \tag{10.5}$$

The width of the region having the field H_1 follows from (3)

$$\Delta y_1 = \rho_1 - \rho_1 \sin \alpha_{11} = \rho_1 \sin \Delta\zeta \tag{10.6}$$

We have now all the parameters describing the first region. $(\rho_1, \Delta y_1, B_1 = B_0 \rho_0 / \rho_1)$ as functions of known quantities ρ_0 and $\Delta\rho_0$.

Consider now an electron emitted at an angle of $2 \Delta\zeta$ which ofter passing the regions with fields B_0 and B_1 enters a region where the field B_2 is still weaker and the radius of curvature $\rho_2 = \rho_1 + \Delta\rho_2$ larger. One can see from Fig. 7b. th at in the same way as for the first region

$$\overline{SO} = \rho_0 = \rho_0 \cos 2 \Delta\zeta + \Delta\rho_1 \cos \alpha_{21} + \Delta\rho_2 \cos \alpha_{22} \tag{10.7}$$

which gives

$$\Delta\rho_2 = \frac{1}{\cos \alpha_{22}} \left[\rho_0 (1 - \cos 2 \Delta\zeta) - \Delta\rho_1 \cos \alpha_{21}\right] \tag{10.8}$$

where α_{21} is obtained from

$$\overline{OA} = \rho_o = \rho_o \sin \alpha_{21} + \rho_o \sin 2\Delta\zeta$$

or

$$\sin \alpha_{21} = 1 - \sin 2\Delta\zeta \qquad (10.9)$$

The angle α_{22} is obtained by putting

$$\Delta y_1 = \rho_1 \sin \alpha_{22} - \rho_1 \sin \alpha_{21} \qquad (10.10)$$

which together with (6) and (9) gives

$$\sin \alpha_{22} = 1 + \sin \Delta\zeta - \sin 2\Delta\zeta \qquad (10.11)$$

The width of the second region Δy_2 is then

$$\Delta y_2 = \rho_2 - \rho_2 \sin \alpha_{22} =$$

$$= \rho_2 (\sin 2\Delta\zeta - \sin\Delta\zeta) \qquad (10.12)$$

Formulae (8) - (12) can be easily generalised, so that for the k-region one would have

$$\rho_k = \rho_o + \sum_{i=1}^{i=k} \Delta\rho_i \qquad (10.13)$$

$$y_k = \rho_o + \sum_{i=1}^{i=k} \Delta y_i \qquad (10.14)$$

$$\Delta\rho_k = \frac{1}{\cos \alpha_{kk}} \left[\rho_o(1 - \cos k\Delta\zeta) - \sum_{i=1}^{i=k-1} \Delta\rho_i \cos \alpha_{ki} \right] \qquad (10.15)$$

$$\Delta y_k = \rho_k \left[\sin k \Delta\zeta - \sin (k-1)\Delta\zeta \right] \qquad (10.16)$$

$$\sin \alpha_{ki} = 1 + \sin (i-1)\Delta\rho - \sin k\Delta\rho \qquad (10.17)$$

$$\sin \alpha_{kk} = 1 + \sin (k-1) \Delta\zeta - \sin k \Delta\zeta \qquad (10.18)$$

This represents a field decreasing stepwise in the direction of y-axis, as shown ofn Fig. 8. By taking very small steps the stepwise field approahes a continuous distribution, also shown on Fig.8. The only discontinuity now remains between the homogenous and inhomogenous region.

The ketron field can be produced with pole-pieces having shapes whown in Fig. 9a and Fig. 9b (14).

A more exact theoretical treatment should include skew rays, which are subject in the inhomogenous part of the field to a slight z-focusing. More detailed theory is given by Pavinskii (15) while Bashilov and Bernotas (16) computed orbits and obtained resolution - transmission characteristics.

Kelman and Yavor (13) quote some results which show that for the same size and resolution, ketron gives several times better transmission than semi-circular spectrometer. Descriptions of ketrons can be also found in references (17) and (18).

Azimuthally varying fields. Sessler was the first to demonstrate (3)that with azimuthally varying field it is possible to eliminate simultaneously the second order aberration terms ζ_r^2 and ζ_z^2 . Further work was then done by Bergkvist and Sessler (4), and by Schmutzler, Daniel and Tauscher (5), who were looking for a practicall solution, which would lead to design of an AVF spectrometer.

Both groups consider a field having a median symmetry plane, containing an optic circle along which the field is constant (B_o). Then field can be presented in the form

$$\frac{B}{z}(\eta,o) = B_o (1 + \alpha_1(\theta)\eta + \alpha_2(\theta)\eta^2 + \alpha_3(\theta)\eta^3 + ...) \qquad (10.19)$$

where the field coefficients $\alpha_n(\theta)$ depend on the azimuthal angle θ in the following way

$$\alpha_n(\theta) = D_{on} + D_{1n} \sin\theta + D_{2n} \cos\theta + D_{3n} \sin 2\theta +$$

$$+ D_{4n} \cos 2\theta + D_{5n} \sin 3\theta + D_{6n} \cos 3\theta \qquad (10.20)$$

For a point source on the optic circle, the radial coordinate of the image point η_f is expressed by

$$\eta_f = \sum_{\mu,\nu} A_{\mu,\nu} \zeta_r^\mu \zeta_z^\nu \qquad (10.21)$$

The values of D_{mn} have to be found which annulate the coefficients A_{o2}, A_{2o} and some higher-ones. There is a wide variety of choices of these coefficients, which give first-order double-focusing, within a wide range of azimuthal angles. Both groups of authors have limited their studies to focusing angles smaller than 2π

Since analytic determination of D_{mn} coefficients, which annulate $A_{\mu\nu}$, would represent a very tedious work, the solutions are found by digital computers. The calculations are facilitated by the fact that for $\mu + \nu = n$, $A_{\mu,\nu}$ is a linear function of D_{mn} and does not depend on higher order α's ($\alpha_{n'}$, where n' > n), which permits to annulate gradually higher and higher orders of $A_{\mu\nu}$.

Bergkvist and Sessler have calculated a field in which the radial aberrations have been eliminated approximately to the sixth order, for a point source, with following characteristics:

$$\Phi_r = 2\Phi_z = 350^o$$
$$R = 0,01\%$$
$$\Omega = 0,9\%$$
$$D = 10$$

Radial magnification: - 0, 81

Axial magnification: - 0, 9

Field coefficients:

$$\alpha_1 = -1,11473 - 0,659 \cos \theta$$
$$\alpha_2 = 1,2230 + 1,1837 \cos \theta$$
$$\alpha_3 = -1,1335 - 1,3550 \cos \theta$$
$$\alpha_4 = 0,8865 + 1,2770 \cos \theta - 0,0066 \sin 2\theta$$
$$\alpha_5 = 0,090 + 0,040 \cos \theta + 0,548 \cos 2\theta$$
$$\alpha_6 = -0,22$$

The above field coefficients $\alpha_n(\theta)$ were obtained with less parameters D_{mn} than available. The authors have shown that the remaining parameters can be used to reduce the aberrations due to cross-terms such as $\tau_o \zeta_z$.

Bergkvist and Sessler have not found a magnet which would produce the field defined above. They suggest that it would be more convenient to start from the beginning with a realistic coil geometry, having some adjustable parameters and then try to find combinations of these parameters, for which the aberrations coefficients are vanishing.

Schmutzler has made calculations (19) for an extended source (0, 25 x 25 mm^2 and R_o = 50 cm) and obtains a resolution of 0.012% at a solid angle of 0.38%, the radial focusing angle being 402° and dispersion D = 10.

The initial theoretical investigations in Berkeley and Heidelberg show that AVFS is a promising type of spectrometer. Much further work will be needed to exploit fully its potentialities.

Table 10.1

POSSIBLE CORRECTOR TYPES

	Spectrometer	
Corrector type	Iron-cored	Iron-free
Electrostatic	Made by Bergkvist (*109*)	Calculated by Lee-Whiting (*112*)
Mag-netic — Iron-free	Made by Shibata *et al.* (*113*)	General theory by Shpalek (*114*)
Mag-netic — Iron	Not reported	Not convenient (Field disturbed)

References for Ch.10

1. K.E.Bergkvist, Ark. Fys. 27, 383 and 439 (1964), also Nucl. Instr.Meth. 43, 170 (1966)

2. R.J.Walen, Nucl.Instr. Meth. 1, 242 (1957)

3. A.M.Sessler, Nucl.Instr.Meth. 23, 165 (1963)

4. K.E.Bergkvist and A.M.Sessler, Nucl.Instr.Meth. 46, 317 (1967)

5. F.Schmutzler, H.Daniel and L.Tauscher, Nucl.Instr.Meth. 35, 175 (1965)

6. F.Schmutzler, Ph. D.Theses, Heidelberg, 1968

7. H.Daniel, P.Jahn, M.Kintze and G.Spannagel, Nucl.Instr.Meth. 35, 171 (1965)

8. B.A.Gumenyuk, G.L.Granberg, G.D.Ivanova, B.G.Peregud and L.A. Cherstov, Nucl. Instr. Meth. 60, 164 (1968)

9. G.Lee-Whiting, Nucl.Instr. Meth. 43, 194 (1966)

10. T.Shibata, S.Taya and Y.Yoshizawa, Nucl. Instr. Meth. 64, 29 (1968)

11. Shpalek, Communicated at Bled Meeting on Beta-ray Spectrometers,1968, Unpublished

12. M.Korsunskii, V.Kelman and B.Petrov, Zhurn. Eksp.Teor. Fiz. 14, 394 (1944)

13. V.M.Kelman and S.Ya.Yavor: "Elektronaya Optika", Akad. Nauk SSSR, Moskva, 1963, p.304-307.

14. B.S.Dzhelepov and S.A. Shestopalova, Izvestia AN SSSR,Ser. Fiz. 20, 328 (1956)

15. P.P.Pavinskii, Izvestia AN SSSR, Ser. Fiz. 18, 175 (1954)

16. A.A.Bashilov and V.I.Bernotas, Izvestia AN SSSR, Ser. Fiz. 18, 192 (1954)

17. B.S.Dzhelepov and A.A.Bashilov, Izvestia AN SSSR, Ser. Fiz. 14, 263 (1950)

18. G.D.Latishev, A.G.Sergeev, E.M.Krisyuk, L.A.Ostretzov, Y.S.Yegorov and N.M.Shirkov, Izvestia AN SSSR, Ser. Fiz. 20, 354 (1956).

Text to Figures. Ch.10

Fig.1.   Principle of multistrip source.

Fig.2.   Bergkvist multistrip source.

Fig.3.   Corrector for semi-circular spectrograph.

Fig.4.   Principle of electro-static corrector.

Fig.5.   Principle of the coil corrector for iron-core spectrometer.

Fig.6.   Geometry of the coil corrector.

Fig.7.   The trajectory in first (a) and second (b) field correcting step.

Fig.8.   Step-wise field for ketron.

Fig.9.   Pole-pieces shape producing ketron.

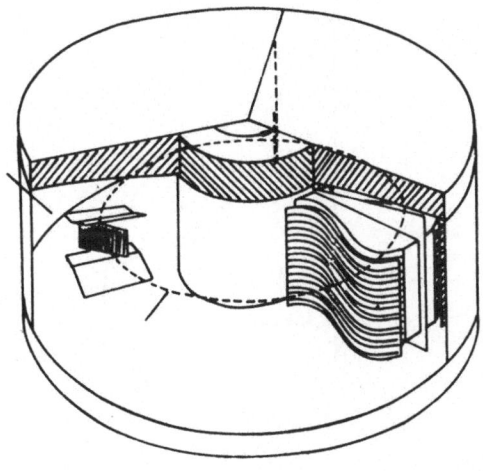

10.1

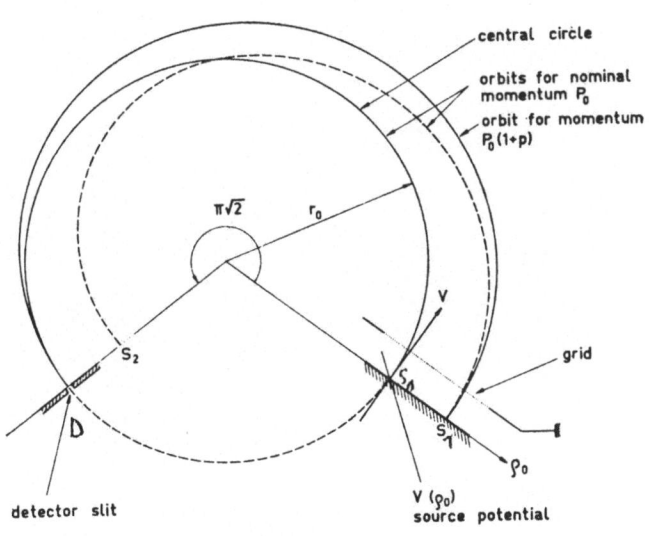

central circle

orbits for nominal momentum P_0

orbit for momentum $P_0(1+p)$

$\pi\sqrt{2}$　　r_0

S_2

grid

D

S

S_1

V

ρ_0

detector slit

$V(\rho_0)$
source potential

10.2

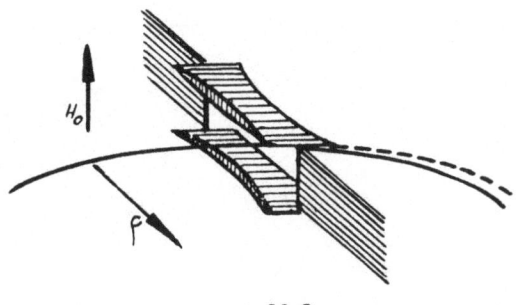

H_0

ρ

10.3

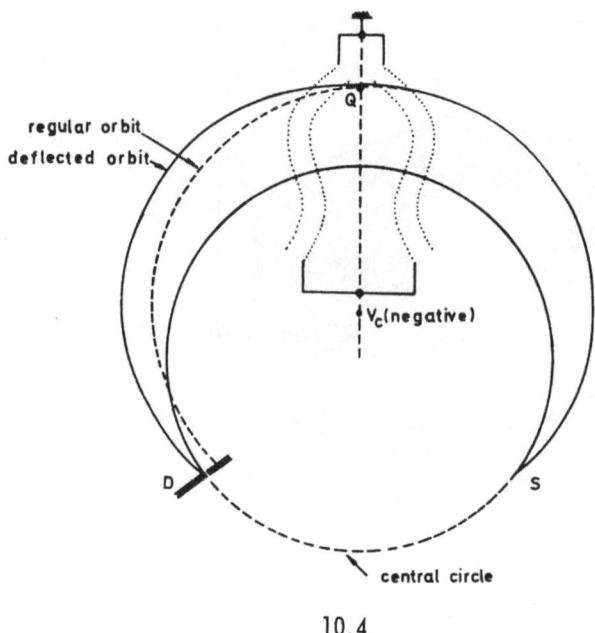

regular orbit
deflected orbit

Q

V_c(negative)

D

S

central circle

10.4

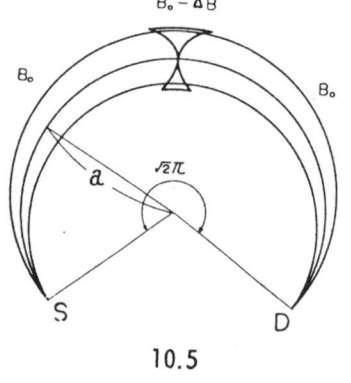

$B_o - \Delta B$

B_o

B_o

a

$\sqrt{2}\pi$

S

D

10.5

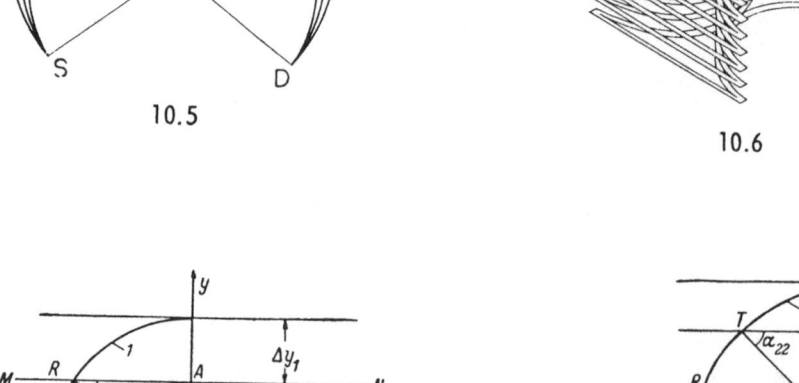

OPTICAL CIRCLE

10.6

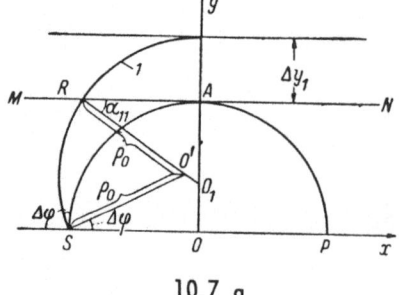

10.7.a

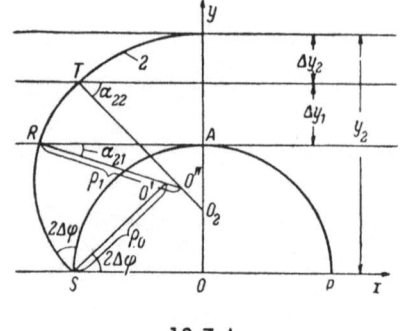

10.7.b

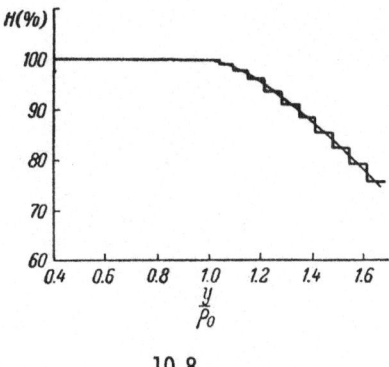

10.8

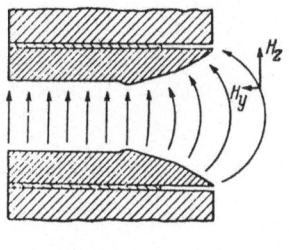

10.9

11. TOROIDAL ("ORANGE") SPECTROMETERS

Orange spectrometers have much higher transmissions than any other type of spectrometer, the values of T ranging between 10-20%. For a given T the resolution is an order of magnitude better than in high transmission lens spectrometers. The resolutions are good, but limited on high R side, due to not yet fully mastered fringing field effects. Orange spectrometers are usefull in the R region from a few pro mille to a few percent. The resolutions beyond 0,1% have remained so far out of reach.

Iron-core orange spectrometers, built so far, usually have 6-8 gaps. Some sector type spectrometers with the geometry similar to a single orange gap have also been constructed. Iron-free orange spectrometers have an order of magnitude greater number of gaps (up to 100).

Most of the orange spectrometers, built so far, are of the symmetrical type, pole profiles on source side being identical to those on detector side. For that reason we shall limit our discussion to symmetrical instruments.

A very detailed theoretical discussion of orange spectrometer was made by Argentine group: Jaffey, Mallman, Suarez-Etchepare and T.Suter (1). It is at the same time one of the most complete theoretical analysis of a magnetic beta-ray spectrometer over made.

In order to illustrate their theoretical approach we shall reproduce their derivation of ideal pole profile and dispersion, taking more simple case of a symmetrical instrument.

Only general conclusions regarding the fringing field effects will be given, since the derivations are lengthy and can be found in their paper.

In principle, spectrometers could be made with electrons describing, inside the magnet, one or more complete loops and then leaving towards the detector. No such spectrometer has been constructed yet, and we shall not discuss it either, reffering the reader to the theoretical study of Argentine group (2). It should be mentioned that special diaphragms are needed to prevent lower energy electrons making several loops to reach the detector and produce the "ghost peaks".

1. Theoretical analysis

Field boundary shape. The source and the detector are placed on
the axis of a torus. Electrons are moving in field-free region untill they reach
the torus, where they are deflected by a field proportional to r^{-1} and after
emerging from it, they are focused into detector on the axis. (Fig.1).

The focusing depends primarily on the shape of the field boundary
at the entry and exit side of the beam. The problem of boundary shape is usually approached in two steps. In first approximation the field is supposed to be
confined within boundaries ànd fringing field effects neglected. Then in next
step, corrections are made for fringing field effects. In this section we shall
consider ideal field, proportional to r^{-1} within boundaries and falling to zero
abruptly outside of them. First, trajectories are studied in the field and then
using appropriate boundary condition, they are extended by straight lines into
focal points.

In the cylindrical system of coordinates, the field can be represented as

$$B_r = 0 \quad B_z = 0 \quad B_\phi = B_o/r$$

where B_o is the field at unit distance from the z-axis.

If the fringing field effects are neglected, electron trajectory remain
in a plane containing z-axis, because B_ϕ is perpendicular to it. The motion
is two-dimensional and can be described by two equations, one of them representing the projection of motion on z-axis.

$$\ddot{z} = (e/m)\,\dot{r}B_\phi \tag{11.1}$$

and another, constancy of velocity in magnetic field

$$(\dot{r})^2 + (\dot{z})^2 = v^2 \tag{11.2}$$

Angle Ψ between velocity vector and positive x-axis can be used as
parameter describing the path. Components of velocity are then expressed by

$$\dot{z} = v \, \cos \Psi \tag{11.3}$$

$$\dot{r} = v \sin \Psi \tag{11.4}$$

Introducing another parameter a as value of r when z = o,
equation (1) can be integrated, giving

$$\dot{z} = \frac{eB_o}{m} \ln \frac{r}{a} \qquad (11.5)$$

Equating (3) and (5)

$$v \cos \psi = \frac{eB_o}{m} \ln \frac{r}{a}$$

$$\ln \frac{r}{a} = - K \cos \psi \qquad (11.6)$$

$$r = a\, e^{-K \cos \psi}$$

with

$$K = - \frac{mv}{eB_o} = \frac{\rho_c}{r}$$

where ρ_c is the radius of curvature of trajectory.

The z-coordinate can be expressed by parameters ψ , a and K
starting from equation

$$z = \int \frac{\dot{z}}{\dot{r}}\, dr + z_o \qquad (11.7)$$

Introducing (3) and (4), (7) becomes

$$z = aK \int \cos \psi\ e^{-K \cos \psi} d\psi + z_o \qquad (11.8)$$

$$= aK\ U(K, \psi) + z_o$$

where $z = z_o$ for $\psi = \pi$ and

$$U(K, \psi) = \int_{\pi}^{\psi} \cos \psi\ e^{-K \cos \psi} d\psi \qquad (11.9)$$

$$= - i J_1(i K)(\pi - \psi) + \sum_{\ell=1}^{\infty} \frac{1}{\ell}\ i^{(\ell-1)} \left[J_{\ell-1}(iK) - J_{\ell+1}(iK) \right] \sin \ell \psi$$

A table of U(K, ψ) developed in series of Bessel functions (9), can
be found in reference (1).

Equations

$$\dot{z} = a\, KU(K, \psi) + z_o \qquad (11.8)$$

$$r = a e^{-K \cos \psi} \qquad (11.9)$$

describe cycloidal loops shown in Fig.2. The form of loops depend on parameter K. Fig.3 shows a family of loops having K = 1 and different values of
parameter a, which determines the height of individual loops in the family.
When dealing with symmetrical orbits as in Fig.3, it is convenient to put
$z_o = 0$.

In a spectrometer with symmetrical entry and exit field boundaries,
the focal distances of source and detector are equal. The boundary is determined by points (Z_e, r_e) at which the straight lines, starting from the source,
tuch a family of curves with a given K (Fig.4). An electron emitted at an angle
Ψ_s will reach field boundary at a point (r_e, z_e) determined by the relation

$$\frac{r_e}{z_e - z_s} = \tan \Psi_s \qquad (11.10)$$

or

$$z_e = z_s + r_e \cot \Psi_s$$

Putting (6) and (8) into (10), parameter a can be expressed in
terms of Ψ_s as

$$a_e = \frac{z_s}{KU(K, \Psi_b) - e^{-K \cos \Psi_s} . ctg \Psi_s} \qquad (11.11)$$

Putting this value of a into relations (6) and (8), following relations
are obtained between boundary points (r_e, z_e) and emission angles Ψ_s

$$z_e = \frac{z_s KU(K, \Psi_s)}{KU(K, \Psi_s) - e^{-K \cos \Psi_s} . \cot \Psi_s} \qquad (11.12a)$$

$$r_e = \frac{z_s e^{-K \cos \Psi_s}}{KU(K, \Psi_s) - e^{-K \cos \Psi_s} . \cot \Psi_s} \qquad (11.12b)$$

The equations (12) describe boundary curve for the simplest case
of symmetric orbits and symmetric boundary shapes with respect to z = 0.
A family of such boundary shapes is shown in Fig.5 for different values of K.

The choice of boundary shape depends on several factors, one of
the most important being the aberrations caused by fringing field.

The relation (12b) can be quite generally written as

$$r_e = z_s f_s (\Psi_s)$$ (11.13)

where f_s represents the profile curve. Then from (10) it follows that

$$z_e = z_s \left[1 + f_s (\Psi_s) \cot \Psi_{s} \right]$$ (11.14)

In the same way, at the exit side

$$r_o = z_f f_f (\Psi_f)$$ (11.15)

$$z_o = z_f \left[1 + f_f (\Psi_f) \cot \Psi_f \right]$$ (11.16)

Dispersion. The usual way to find the dispersion is to consider the difference in the position of focus for two particles emitted in the same direction, but having a small difference in momentum. We shall assume that particle with momentum p describes a symmetric trajectory, as shown in Fig.6. The source position z_s, emission angle Ψ_s and the field entering point (z_e, r_e) are connected with corresponding values at the focusing side by the relations

$$z_f = - z_s$$
$$r_o = r_e$$
$$z_o = - z_e$$ (11.17)
$$\Psi_f = 2\pi - \Psi_s$$

The particle having the momentum $p + \delta p$ starts from z_s at angle Ψ_s, enters the field at $(z_e r_e)$, but leaves the field at $(z_o + \delta z_o, r_o + \delta r_o)$ and is focused at $z_f + \delta z_f$. Since

$$\frac{\delta p}{p} = \frac{\delta K}{K}$$ (11.18)

particle with momentum $p + dp$ will have the trajectory caracterised by $K + \delta K$. This trajectory can be determined as a variation of the trajectory of a particle with momentum p.

The particle emitted at z_s, entering the field at (r_e, z_e) describes a trajectory determined by relations

$$r = r_e e^{-K (\cos \Psi - \cos \Psi_s)}$$ (11.19)

$$z = z_e + r_e\, e^{K\cos\,\Psi_s}\, K\, \left[U(K,\,\Psi) - U(K,\,\Psi_s)\right] \qquad (11.20)$$

$$F_o = \left[U(K,\,\Psi) - U(K,\,\Psi_s)\right] \qquad (11.21)$$

The varied trajectory is obtained by varying K and Ψ. Differentiating (19) and (20)

$$\delta r = r_e\, e^{-K(\cos\Psi\,-\cos\,\Psi_s)}\left[K\sin\,\Psi\delta\Psi - (\cos\,\Psi - \cos\,\Psi_s)\,\delta K\right]$$

$$\delta z = r_e\, e^{K\cos\,\Psi_s}\left\{(\cos\,\Psi_s\, F_o + \frac{\delta\dot F_o}{\delta K})\,\delta K + K\cos\,\Psi\, e^{-K\cos\Psi}\delta\Psi\right\} \quad (11.23)$$

There we have used the relation

$$dU(K,\,\Psi) = \cos\,\Psi\, e^{-K\cos\Psi}\, d\Psi \qquad (11.24)$$

The varied trajectory meets the exit boundary when $\Psi = \Psi_f$. For the points on the exit boundary δr and δz become

$$\delta r = r_e\, e^{-K(\cos\,\Psi_f-\cos\,\Psi_s)}\left[K\sin\,\Psi_f\,\delta\Psi_f - (\cos\,\Psi_f - \cos\,\Psi_s)\,\delta K\right]$$

$$= r_o\left[K\sin\,\Psi_f\delta\Psi_f - (\cos\,\Psi_f - \cos\,\Psi_s)\,\delta K\right] \qquad (11.25)$$

$$\delta z = r_o\left\{e^{K\cos\,\Psi_f}(\cos\,\Psi_s\cdot F + \frac{\partial F}{\partial K}\delta K + K\cos\,\Psi_f\delta\Psi_f\right\} \qquad (11.26)$$

$$F = K\left[U(K,\,\Psi_f) - U(K,\,\Psi_s)\right] \qquad (11.27)$$

These relations are simplified when account is taken of left-right symmetry, which requires that following should hold

$$\cos\,\Psi_f = \cos\,\Psi_s$$

$$\sin\,\Psi_f = -\sin\,\Psi_s$$

$$U(K,\,\Psi_f) = -U(K,\,\Psi_s) \qquad (11.28)$$

$$F = -2KU\,(K,\,\Psi_s) = F_s$$

and (25) and (26) become

$$\delta r = r_o\, K\sin\,\Psi_f\delta\Psi_f \qquad (11.29)$$

$$\delta z = r_o \left\{ e^{K\cos \Psi_f} (\cos \Psi_s F_s + \frac{\partial F_s}{\partial K}) \delta K + K\cos \Psi_f \delta \Psi_f \right\} \tag{11.30}$$

The varied exit point $(z_o + \delta_o, \ r_o + \delta r_o)$, in which we are interested, can be reached from focus z_f by changing Ψ_f to $\Psi_f + \delta\theta$. Differentiating (15) with respect to Ψ_f, and putting $\Psi_f = \theta$

$$\delta r_o = z_f \frac{d_f(\Psi_f)}{d\Psi_f} \quad \delta\theta = r_o \frac{f_f'(\Psi_f)}{f_f(\Psi_f)} \delta\theta \tag{11.31}$$

$$= r_o \Lambda_f \delta\theta \quad \Lambda_f = \frac{r_f'(\Psi_f)}{f_f(\Psi_f)} \tag{11.32}$$

$$\delta z_o = z_f \left| f_f'(\Psi_f) \cot \Psi_f - \frac{f_f(\Psi_f)}{\sin^2 \Psi_f} \right| \delta\theta$$

$$= r_o \left[\Lambda_f \cot \Psi_f - \frac{1}{\sin^2 \Psi_f} \right] \delta\theta \tag{11.33}$$

The point where the varied trajectory meets the exit boundary will be the point $(z_o + \delta z_o, \ r_o + \delta r_o)$, if δr and δz are equal to δr_o and δz_o respectively. When (29) is equalized to (31), and (30), to (33), we obtain following values for $\delta\theta$ and $\delta\Psi_f$.

$$\delta\theta = \frac{1}{\Lambda_f} K \sin \Psi_f \delta\Psi_f = \frac{1}{\Lambda_f} K \sin \Psi_s \delta\Psi_s \tag{11.34}$$

$$\delta\Psi_f = - \Lambda_f \sin \Psi_f e^{K\cos \Psi_f} F_2 \frac{\delta K}{K} \tag{11.35}$$

$$F_2 = F \cos \Psi_s + \frac{\partial F}{\partial K} \tag{11.36}$$

The values of $\delta\theta$ and $\delta\Psi_f$ which we have just found can be connected to δz_f, which we look for, by starting from (10)

$$z_o = z_f + r_o \cot \Psi_f$$

which, differentiated becomes

$$\delta z_o = \delta z_f + \cot \Psi_f \cdot r_o + \frac{r_o}{\sin^2 \Psi_f} \delta\Psi_f \tag{11.37}$$

Introducing (31) and (33) we obtain

$$\delta z_f = \frac{r_o}{\sin^2 \Psi_f} (\delta \Psi_f - \delta \theta) \tag{11.38}$$

When we use (34) and (35) δz_f becomes

$$\delta z_f = \frac{r_o}{\sin \Psi_f} (\Lambda - K \sin \Psi_f) e^{K \cos \Psi_s} F_2 \frac{\delta K}{K} \tag{11.39}$$

The difference $\Lambda - K \sin \Psi_f$ can be derived from the equation of exit boundary which follows from (12b) and (15)

$$f_f (\Psi_f) = \frac{1}{K e^{-K \cos \Psi_f} U(K, \Psi_f) - \cot \Psi_f} \tag{11.40}$$

which differentiated gives

$$\Lambda - K \sin \Psi_f = - \frac{r_o}{z_f} \frac{1}{\sin^2 \Psi_f} \tag{11.41}$$

Introduced into (39), δz_f becomes

$$\delta z_f = \frac{r_o^2}{z_f} \cdot \frac{1}{\sin^3 \Psi_f} e^{K \cos \Psi_s} F_2 \frac{\delta K}{K}$$

$$= \frac{r_e^2}{z_s} \cdot \frac{1}{\sin^3 \Psi_f} \cdot e^{K \cos \Psi_s} F_2 \frac{\delta K}{K} \tag{11.42}$$

$$F_2 = - 2 U(K, \Psi_s) \left[K \cos \Psi_s - 1 \right] - 2 K \frac{\partial U(K, \Psi_s)}{\partial K} \tag{11.43}$$

Dispersion D defined as

$$D = - \frac{dz_f}{dp/p} = - \frac{dz_f}{dk/k}$$

becomes

$$D = - \frac{r_e^2}{z_s} \cdot \frac{1}{\sin^3 \Psi_f} e^{-K \cos \Psi_s} F_2 \tag{11.44}$$

It is convenient to introduce a function G defined by

$$D = 2 z_s G \tag{11.45}$$

and given explicitly by

$$G = - \frac{\delta z_f}{2 z_s} \cdot \frac{1}{\delta K/K} = - \frac{1}{2} \frac{r_e^2}{z_s^2} \frac{e^{K\cos \Psi_s}}{\sin^3 \Psi_f} \cdot F_2 \tag{11.46}$$

The function G is shown in Fig. 7, expressed in units corresponding to $G = 1$ as dispersion factor of semicircular spectrometer. It can be seen that dispersion is greater for smaller Ψ_s and K. Since in a given spectrometer K is fixed, while a range of entrance angles Ψ_s is accepted, the effective dispersion is obtained by averaging the corresponding G values.

Fringing field effects. Toroidal spectrometers so far built have 6-8 gaps in iron instruments and 36-100 in iron-free ones. The gaps deform the field inside and create a field outside of pole-pieces or coils. The field distorsion due to final openings of the gaps contribute in several ways to the focusing properties. Only brief discussion of fringing field effects is given below. A full analysis can be found in the paper of Yaffe et al (1).

The fringing field acting on a particle, which started from the focus on the z-axis, can be decomposed in three components, $\vec{B}_\phi$-normal to ϕ plane containing the trajectory, $\vec{B}_t$ along the trajectory and $\vec{B}_\perp$ normal to it, both lying in the ϕ plane (Fig. 8). The field component parallel to the particle momentum does not contribute to its movemovent, so that only $\vec{B}_\phi$ and $\vec{B}_\perp$ have to be taken into account.

The intensity of the $\vec{B}_\phi$ component along the trajectory of the particle is shown in Fig. 9. As particle leaves the source the field rises from zero linearly (very approximately) and on entering the gap, the intensity gradually approaches the undistorted value. At a distance from the edge roughly equal to the gap width, the effect of fringing field is reduced practically to zero. The $\vec{B}_\phi$ component produces the trajectory deflection of the same kind as interior field. It's contribution to focusing is equivalent to having a larger field region. The correction, however, is not so simple to make because $\vec{B}_\phi$ varies along a single line of force as well as along the profile. For symmetry reasons $\vec{B}_\phi$ is maximum in the median plane, where other two components are equal to zero. The correction along the profile for the effect of $\vec{B}_\phi$ in median plane is relatively easier to make, than for Φ - variation of $\vec{B}_\Phi$ acros the gap.

The component $\vec{B}_\perp$ which is contained in the Φ_s -plane and at the same time normal to the momentum of the particle, produces a force perpendicular to Φ_s-plane, moving the particle out of that plane. The magnitude of this lateral force is equal to $evB_\perp = evB_t \tan \mu$, where μ is the (smaller) angle between the normal to the profile and the trajectory, since the lines of force in the fringing field are contained within the planes normal to the profile. The lateral force is equal to zero in the median plane and has opposite signs on the two different sides of that plane. This is equivalent to the action of a cylindrical lens, having the axis in the median plane and following the profile.

The cylindrical lens is converging to median plane, or diverging from it, depending on the sign of tan μ. The lens is converging on the entrance side if the profile distance decreases as Ψ_s increases (and opposite for exit side).

The lateral deflection, produced by lens effect may lead to the trajectory missing the detector slit and thus reducing the transmission.

In chapter 16 it is shown that the focal length f of a cylindrical lens is given by (16.9)

$$f = \rho_c \cot \mu$$

where ρ_c is radius of curvature of the trajectory. In the cylindrical lens the image of a point is a line paralel to lens axis. The Φ-focusing is thus disturbed, and a point source broadens into a disk image, the radius of which increases with angle μ, gap angular opening ζ and depends also on the interval of accepted Ψ values and K-parameter.

The lens effect is diminished when the angle μ is reduced and the ideal solution would be to have profiles of circular shape with source and detector in centers of circles. So far most of the spectrometers were designed following this requirement.

All fringing field effects are reduced when smaller angular gap width ζ is used. This is easier to achive in an iron-free, than in an iron spectrometer.

Resolution. There are three important contribution to the resolution of a toroidal spectrometer, one comes from the geometry and dispersion of the instrument, another derives from fringing field effect, and a third may appear when orange slices do not follow perfect cylindrical symmetry.

1. The first one can be calculated, when relevant parameters are known. We shall only consider the simpler case of one-dimensional source and detector placed along z-axis. Since an axial source represent a projection of a radial source, it is easy to connect one with another.

In the symmetrical spectrometer the magnification is equal to 1, and the length of the image δz_f is equal to the source length δz_s when of course fringing field is neglected. The best theoretical ratio of transmission to resolution would then be obtained by having the detector slit width d equal to δz_s. The resolution R is then simply

$$R = \delta z_s / 2 z_s G(\Psi_{sh}) \qquad (11.47)$$

where G is given by (46), and Ψ_{sh} is a mean angle connected with minimum and maximum aperture angles Ψ_{sm} and Ψ_{sM} by the relation

$$\cos \Psi_{sh} = \frac{1}{2} (\cos \Psi_{sm} + \cos \Psi_{sM}) \qquad (11.48)$$

2. The effects of fringing field on the resolution can be divided in two categories, one comprising the additional deflection $\Delta\Psi_a$ due to fringing field and another the lens effect.

a) The additional deflection $\Delta\Psi_a$ may vary along the gap and across the gap. The additional deflection along the profile for trajectories lying in the median plane depends on source profile geometry and the geometry of fringing field. It should be remembered that the gap-width changes with Ψ_s, being larger for smaller values of Ψ_s. The resolution depends on the total deflection $\Delta\Psi_i + \Delta\Psi_a$, where $\Delta\Psi_i$ is the interior deflection. The total deflection is a function of Ψ_s, being larger for smaller values of Ψ_s. In general the variation of $\Delta\Psi_a$ along the profile contribute less to the line-width, than the change of ratio $\Delta\Psi_a / \Delta\Psi_i$. The variation of $\Delta\Psi_a$ is generally smaller than variation of $\Delta\Psi_i$ as Ψ_s is increased. The ratio $\Delta\Psi_a / \Delta\Psi_s$ is smaller at smaller values of Ψ_s. If the profile had the ideal form, the particles emitted at larger angles would be more deflected. This requires a correction of the line profile, which is usually made in semi-empirical way. More perfect correction would cause a smaller contribution of $\Delta\Psi_a$ to the line width.

The fringing field structure changes across the gap and a particle moving in the median plane experiences a different fringing field, than does a particle moving along the gap sides. This effect is more difficult to analize and correct for. The measurement (3), which is reported further bellow, shows that the deflection off median plane is larger. An approximate analysis of Jaffe et al (1) shows that this effect would not impair the resolution only at very small gap openings. At $\Psi = 3^0$ it is negligible, but has to be taken into account at $\Psi = 10^0$.

b) The lens effect broadens the image, requiring a larger detector slit if transmission is not to be sacrified. Larger detector slit causes an increase of line width.

3. The cylindrical symmetry of orange slices represents a design and construction task, more difficult to achieve in case of iron-core than iron-free spectrometers. The resolution of one gap is generally better than the resolution of all gaps of an iron-core spectrometer used simultaneously.

Transmission. The parameters defining the part of the beam accepted by entrance baffles are

ζ = angular aperture of the gaps

Ψ_{sm} = the smallest emission angle accepted

Ψ_{sM} = the largest emission angle accepted

If there are j gaps, the acceptance solid angle Ω is given by relation

$$\Omega = j \int_{sm}^{sM} \zeta \sin\Psi_s \, d\Psi_s = j\zeta \left| \cos\Psi_{sm} - \cos\Psi_{sM} \right| \tag{11.49}$$

The effective transmission is smaller than $\Omega/4\pi$, due to two main causes:

1. When sources are extended, there are trajectories which are not contained in a ϕ-plane. The particle then has a p_ϕ-component of the momentum, which produces a rotation around z-axis. Some of these particles may hit the pole-face or exit baffles (if their angular opening is smaller than ζ) and thus be lost from the beam.

2. Lens effect produces a p_ϕ-component which:

a) can lead to the same kind of loss collision with pole-face, as above,

b) may make the image size larger than detector slit and thus cut--off a part of the beam.

Both of these effects are more serious in case of iron-core spectrometers which have iron polefaces, than in the case of iron-free ones, where coils are hollow, and particles with p_ϕ-component may pass from one gap to another.

The spectrometers

Most of the development of iron-core "orange" was made in Danmark. After a preliminary study (4), Nielsen and Kofoed-Hansen (5) made a six-gap spectrometer, and Nielsen (6) later constructed an enlarged version. Bisgard (7) also constructed a larger version.

Iron-core multigap spectrometers were also constructed by Mallmann (8) and by Spoelstra and Rautenback (9). Dzhelepov, Prihodceva, Tishkin and Shishelov (10) built a coincence spectrometer consisting of two six-gap "oranges".

Iron-free "oranges" were first developed in Moscow, where four apparatus were reported so far; one by Vladimirsky and collaborators (11), two by Tretyakov et al. (13) and one by Burgov and collaborators (14). Double orange made in Argonne by Freedman et al. (15) represents one of the most ambitious beta-spectrometers, with highest performances. Excellent performances were also achieved by Gasior (16) and by Moll (18). An iron-free orange was also constructed by Bartis (17).

Quite a few single-gap iron-cored spectrometers were made, based on the sector boundary shape of multi-gap instrument (4, 19-25).

Szalay and Berenyi have constracted an "orange" spectrometer with uniform magnetic field (26).

We shall give below a short description of Aarhus iron-core spectrometer and Argonne iron-free double spectrometer.

Aarhus iron-core spectrometer. Bisgard (3) has given a detailed description of the characterists of an iron-core toroidal spectrometer built in Aarhus, closely following the original Copenhagen model (5-6), exept for size, which was doubled. The description can serve as a concrete illustration of general discussion given above. A summary of Bisgard´s data is given below.

General characteristics and parameters of the spectrometer are the following:

- Entrance and exit profiles symmetrical
- Number of gaps $j = 6$
- Angular aperture of the gaps $\zeta = 20^{o}$
- Smallest emission angle $\Psi_{sm} \approx 50^{o}$
- Largest emission angle $\Psi_{sM} = 107^{o}$
- Focal distances $z_{s} = - z_{f} = 19$ cm
- Mean profile-focus distance ≈ 18 cm
- Energy range 0, 005 - 5, 0 Mev
- Detector: antracen crystal

Axial and radial sections of the spectrometer are shown in Figs. 10 and 11. The vacuum chamber is 51 cm high with a diameter of 120 cm and contains a magnet which has a total weight of 1000 kg (coils 300 kg and iron parts 700 kg). The maximum power consumption of each coil is only 225 watts, which is easily taken care of by watter-cooling.

Contrary to most of the other spectrometer types, the toroidal coils are inside the vaccuum chamber. The electrical insulator has to be properly chosen and made, not to impair the high vacuum. Aarhus spectrometer coils were impregnated with zincoxide mixed with transformer varnish and probably because of incomplete hardening, the vacuum properties are not very satisfactory

The profiles were empirically corrected for the fringing field effects. The initial form was circular, with the radius of 172 cm. The gap was then divided into small sections by corresponding set of small baffles Then in each region current is measured, which is bringing to focus the same internal conversion line. Higher current was needed in the center of the gap than at the sides, and focusing current is especially reduced at the side of high entrance angles Ψ_s, where the influence of fringing field is greater. Profiles were then mashined until the focusing current along the opening was constant within 0,15% or less.Figure 12 shows the initial and final profiles.

Focusing properties were studied across the gap and it was found that:

- The field near pole plates is about 0,4% stronger than in the median plane. This is due to non-negligible radial opening of the gap.

- Transmission is maximum in median plane and falls off near pole--plates as shown in Fig.13. The transmission loss near pole-plates is partly due to lens effect and partly to the fact that the field is 0,4% stronger than in the median plane.

- For the same reasons the resolution is also best in the median, plane, as shown in Fig.13.

When all the six gaps are used the lack of symmetry is reflected in the fact that it is impossible to achieve the same high resolution limit as obtained with one gap. At transmissions over 10% the resolution for six gaps is about the same as for single one but when transmission decreases to lower limit, the resolution is 0,85% for all gaps and 0,5% for a single gap (when the whole gap is open and contributions of source and detector slit negligible).

The performance of Aarhus spectrometer is represented by T-R curve, for a source of 0.12 x 0,5 cm^2, shown of Fig.14. The curve can be described by the relation

$$T^2 = 1,25 R - 0,0075$$

Argonne spectrometer. Two iron-free spectrometers mounted "back-to-back", have each 100 coils of 1 meter over-all diameter. The gaps are then relatively smaller, minimising the fringing field effects. The acceptance angle is 30°-70°, corresponding to a solid angle of 20%. Source detector distance is 67 cm. The maximum focusing energy 3.5 MeV. The interference between the two spectrometers is negligible.

The performances are the following

Source diam., cm	0,3	0,3	0,6	0,1	0,1
T, %	19	2,8	16	13	1,6
R, %	0,39	0,21	0,56	0,30	0,13

The scheme of the spectrometer is shown in Fig.15. The value of K–parameter is 0,59.

Multiloop iron–free toroidal spectrometer in Moscow. Tretjakov, one of the pioneers of iron–free toroidal spectrometers, built recently an instrument with novel design features (27–28). It can be seen from Fig.5 that for K = 0.74 and $\Psi = 90^{\circ} \pm 5^{\circ}$, the profile curve becomes a straight line parallel to the z–axis. The coil can have rectangular shape, which simplifies the construction and eliminates the need for laborious profile shaping. Such geometry has another two advantages: (a) the magnetic field of rectilinear conductors can be treated analytically, and (b) the coil can be long to allow several electron loops, which improves the resolution and reduces the background. The price payed for those advantages is almost an order of magnitude lower transmission, than normally attained in orange spectrometers. The reason is that not only Ψ –opening is limited to about 10°, but also Φ –opening may be reduced to about one third of 2π, when the source is placed along the axis in order to reduce serious absorption and back–scattering in the source–holder.

The general characteristic of the spectrometer shown on Fig.16 are the following:

- Number of electron loops = 4
- Source – detector distance = 230 cm
- The torus is made from 72 rectangular loops
- Distance of inner portion of rect. loops from
  z–axis = 21 cm
- Distance of outer straight loop portion from
  z–axis = 60 cm

- Maximum distance of mean electron trajectory

from z-axis = 45 cm

The inner straight portions of rectangular loops are made of copper tubes, having outside diameter of 6 mm and 1 mm thick walls. The other three sides of loops, which are not exposed to the electron beam are made of larger tubes.

The efficiency of cooling of the inner tubes limit the maximum energy which can be measured. The author states that by cooling all 72 loops in parallel by distilled water, the electrons of 1-1.5 MeV could be measured.

The performances are illustrated by the following data.

Source size mm^2	T %	R %	L %
0.5 x 10	0.3	0.03	1.5×10^{-4}
0.9 x 10	0.38	0.04	3.4×10^{-4}
2 x 10	0.47	0.07	9.4×10^{-4}
3 x 10	0.47	0.1	1.4×10^{-3}
0.9 x 16	0.22	0.04	3.2×10^{-4}
2 x 16	0.31	0.07	1×10^{-3}
3 x 16	0.35	0.1	1.7×10^{-3}

The luminosity - resolution curve is one of the best ever achieved by any type of beta ray spectrometers.

References for Chapter 11

1. A.H.Jaffey, C.A.Mallman, J.Suarez-Etchepare and T.Suter, ANL–6222.

2. L.Lagata, C.A.Mallman, C.Molina-Vedia, J.Peyre and J.Suarez-Etche-pare, Proceedings of Second UN Geneva Conference, .1958, Vol.14, p.445.

3. K.M.Bisgard, Nucl.Instr.Meth. 22, 221 (1963)

4. O.Kofoed-Hansen, J.Linhard and O.B.Nielsen, Mat.Fys.Medd.Dan.Vid. Selsk, 25 (No. 16) (1950)

5. O.B.Nielsen and O.Kofoed-Hansen, Mat.Fys.Medd.Dan.Vid.Selsk. 29, (No.6) (1955)

6. O.B.Nielsen, Nuclear Phys. 12, 389 (1959)

7. K.M.Bisgard, Nucl. Instr. Meth. 22, 221 (1963)

8. C.A.Mallman, Bull. 34th Meeting of Argentine Phys. Soc. (Sept.1959)

9. B.Spoelstra and W.L.Rautenback Nucl. Instr.Meth. 66, 336 (1968)

10. B.S.Dzhelepov, V.P.Prihodceva, P.A.Tishkin and I.A.Shishelov.Izvestia AN SSSR, Ser. Fiz. 29, 2157 ('965)

11. V.V.Vladimirski, E.K.Tarasov and Yu. V.Trebukhovsky, Prib.Tekn. Eksp. No, 1, 13 (1956)

12. E.F.Tretyakov, L.L.Goldin and G.J.Grishuk, Prib.Tekn.Eksp. No.6, 22 (1957)

13. E.F.Tretyakov, L.N.Kondratyev, G.J.Grishuk, G.J.Novikov and L.L. Goldin, Izvestia AN SSSR, 26, 1470 (1962)

14. N.A.Burgov, A.V.Davydov and G.R.Kartashov, Nucl.Instr.Meth. 12, 316 (1961)

15. M.S.Freedman, F.Wagner, Jr., F.T.Porter, J.Terandy and P.P.Day, Nucl.Instr.Meth. 8, 255 (1960)

16. M.Gasior, Postepy Techniki Jadrowej, 9–10, 859 (1964)

17. F.J.Bartis, Nucl.Instr.Meth. 44, 125 (1966)

18. E.Moll and E.Kankeleit, Nukleonik, 7, 4, 180 (1965)

19. J.Moreau, J.Phys. Rad. 15, 776 (1954)

20. T.Huns, J.H.Bjerregaard and B.Elbek, Dan. Mat. Fys. Medd 30, No.17 (1956)

21. M.F.Shea, R.L.Water and W.C.Miller, Bull. Am.Phys. Soc.5, 226 (1960)

22. E.M.Bernstein and R.Graetzer, Phys. Rev.119, 1321 (1960)

23. H. Van Krugten and E. W. Koopmann, Nucl. Instr. Meth. 71, 280 (1969)

24. K. M. Bisgard, Nucl. Instr. Meth. 29, 213 (1964)

25. A. J. Armini, R. M. Polichar and J. W. Sunier, Nucl. Instr. Meth. 48, 309 (1967)

26. A. Szalay and D. Berényi, Acta Physica Acad. Sci. Hung. 10, 1 (1959)

27. E. K. Tarasov, E. F. Tretiakov, Report of The Institute for Theoretical and Experimental Physics, No. 18, Moscow 1973 (in russian)

28. E. F. Tretiakov, Report of The Institute for Theoretical and Experimental Physics, No. 65, Moscow 1973 (in russian)

Text to Figures. Ch.11

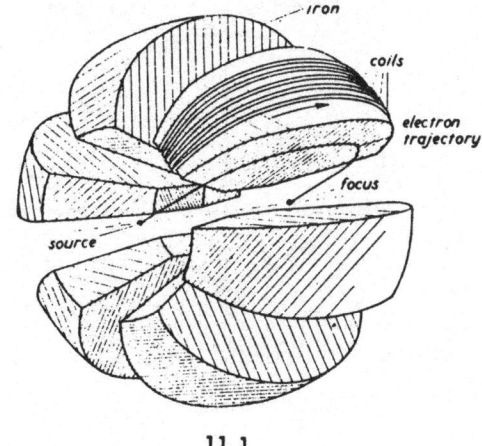

11.1

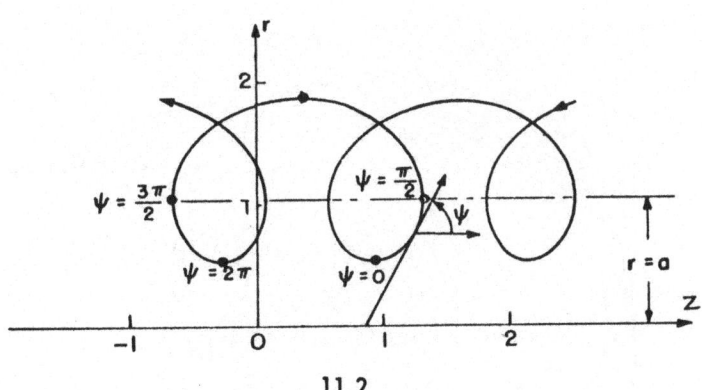

11.2

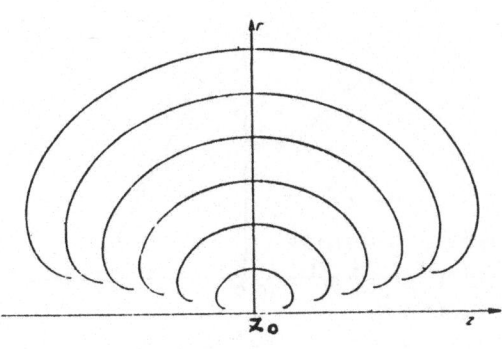

11.3

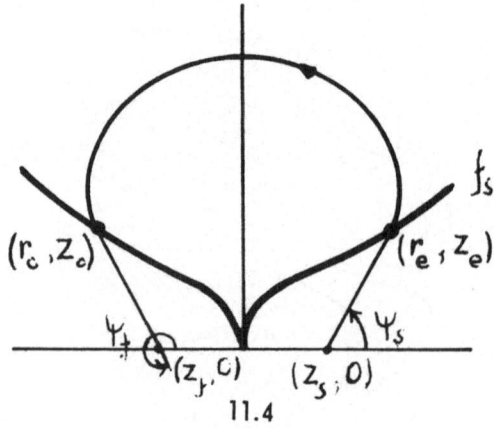

11.4

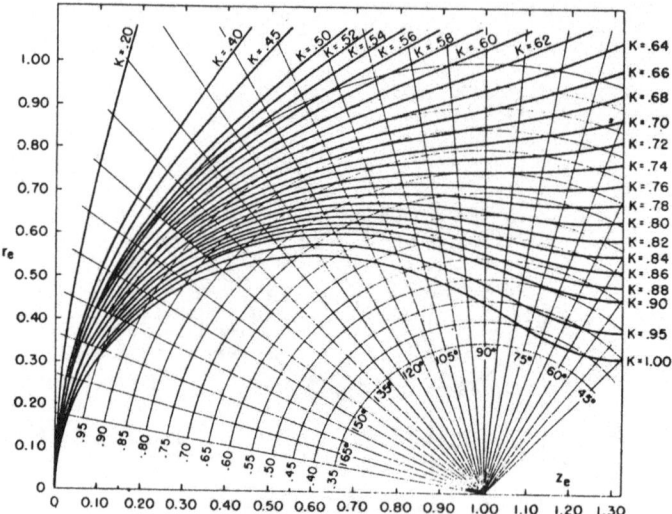

11.5

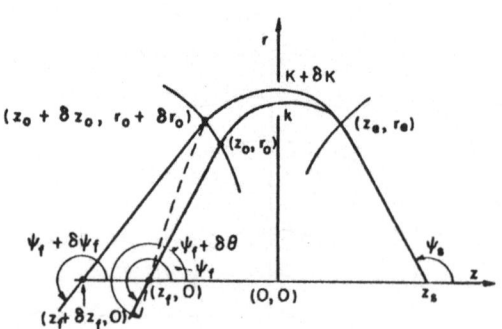

11.6

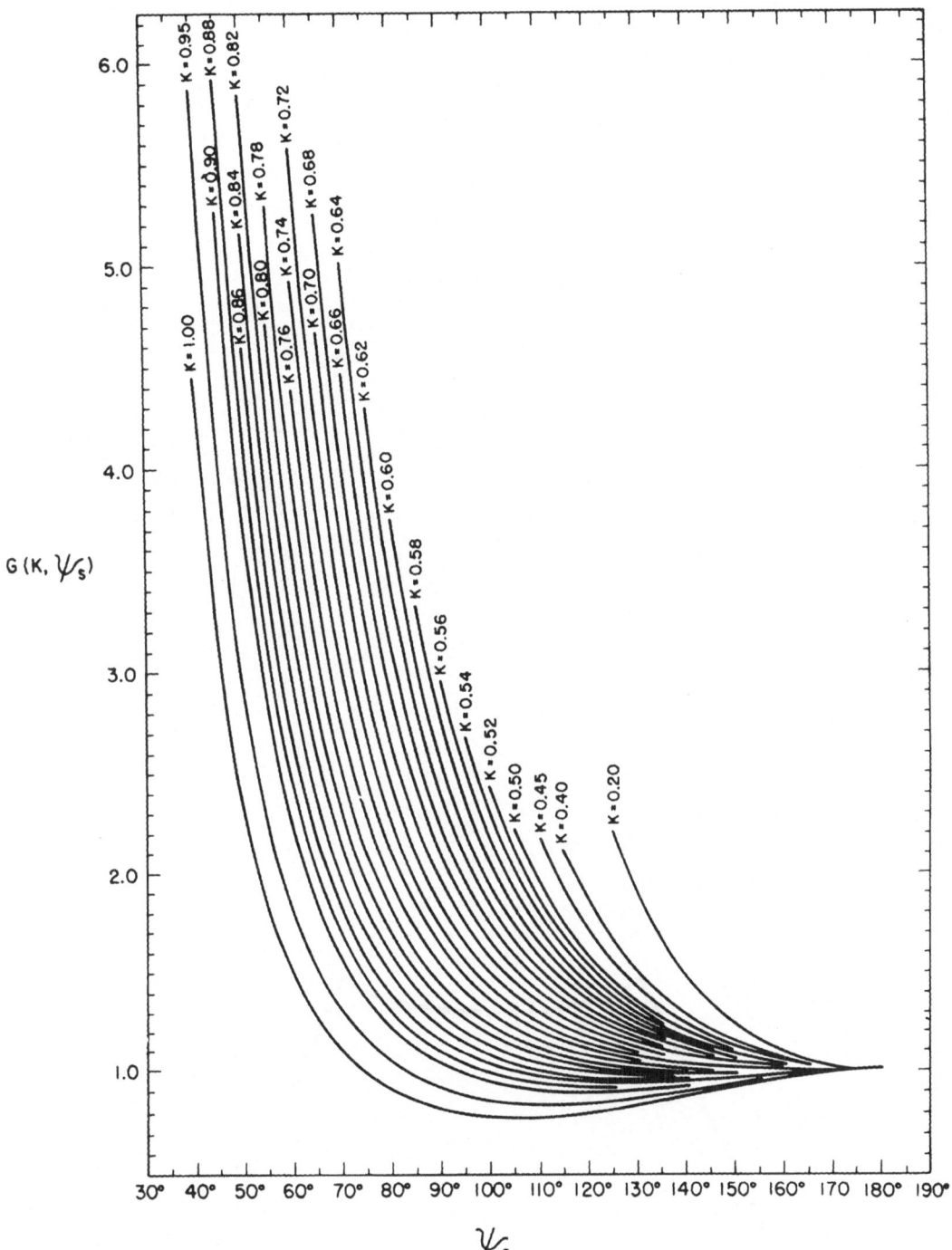

G(K, ψ_s)

ψ_s

11.7

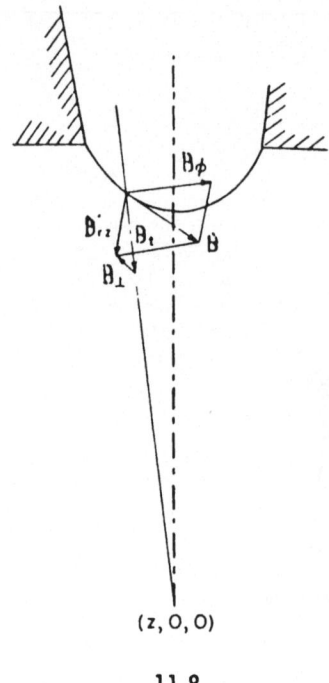

$(z, 0, 0)$

11.8

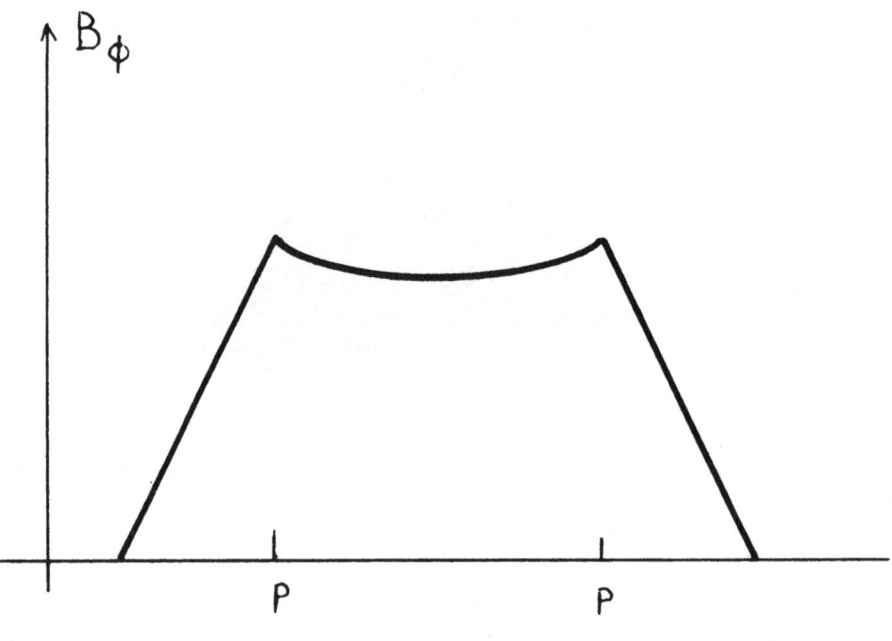

11.9

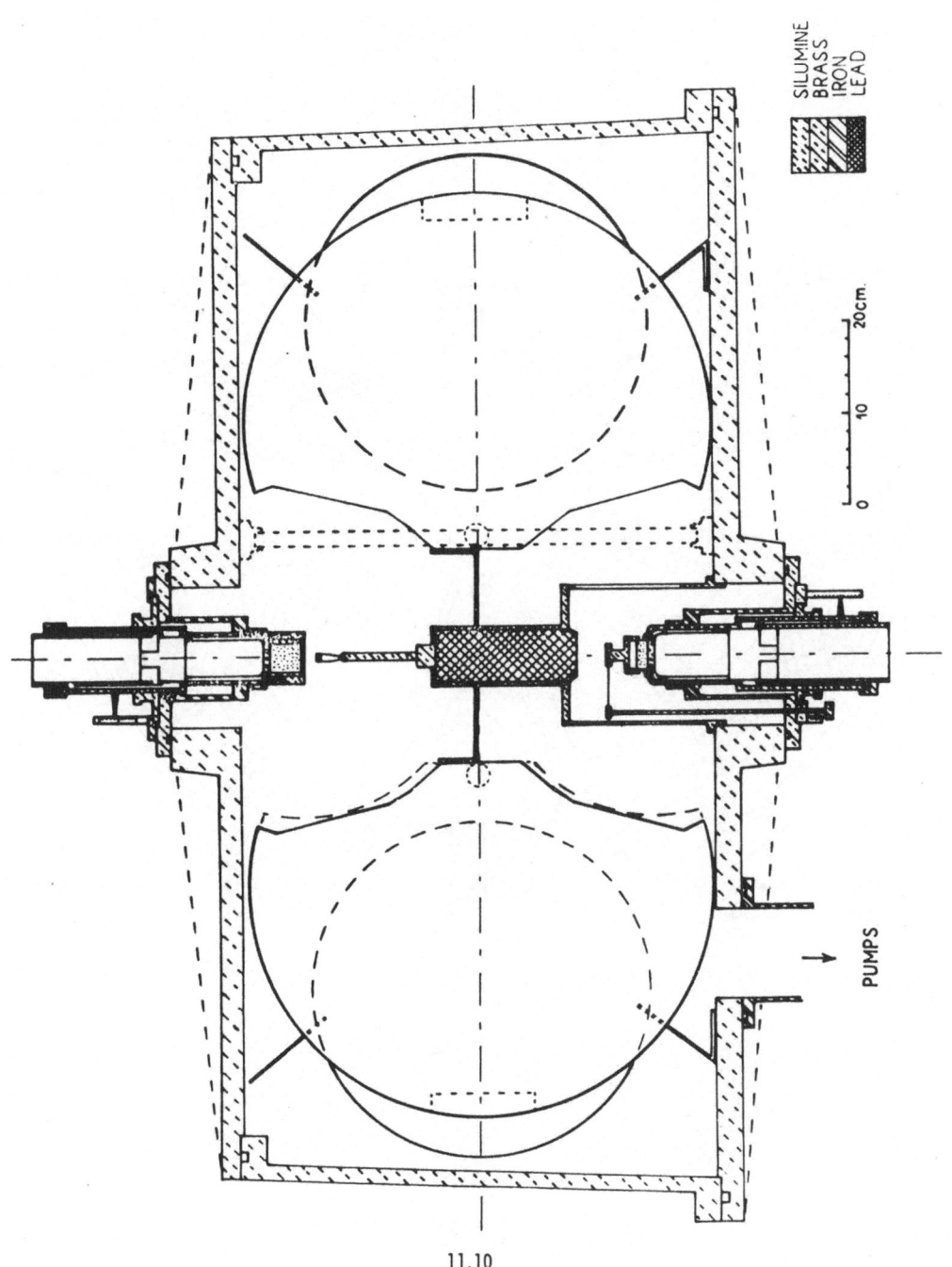

SILUMINE
BRASS
IRON
LEAD

20cm.

10

0

PUMPS

11.10

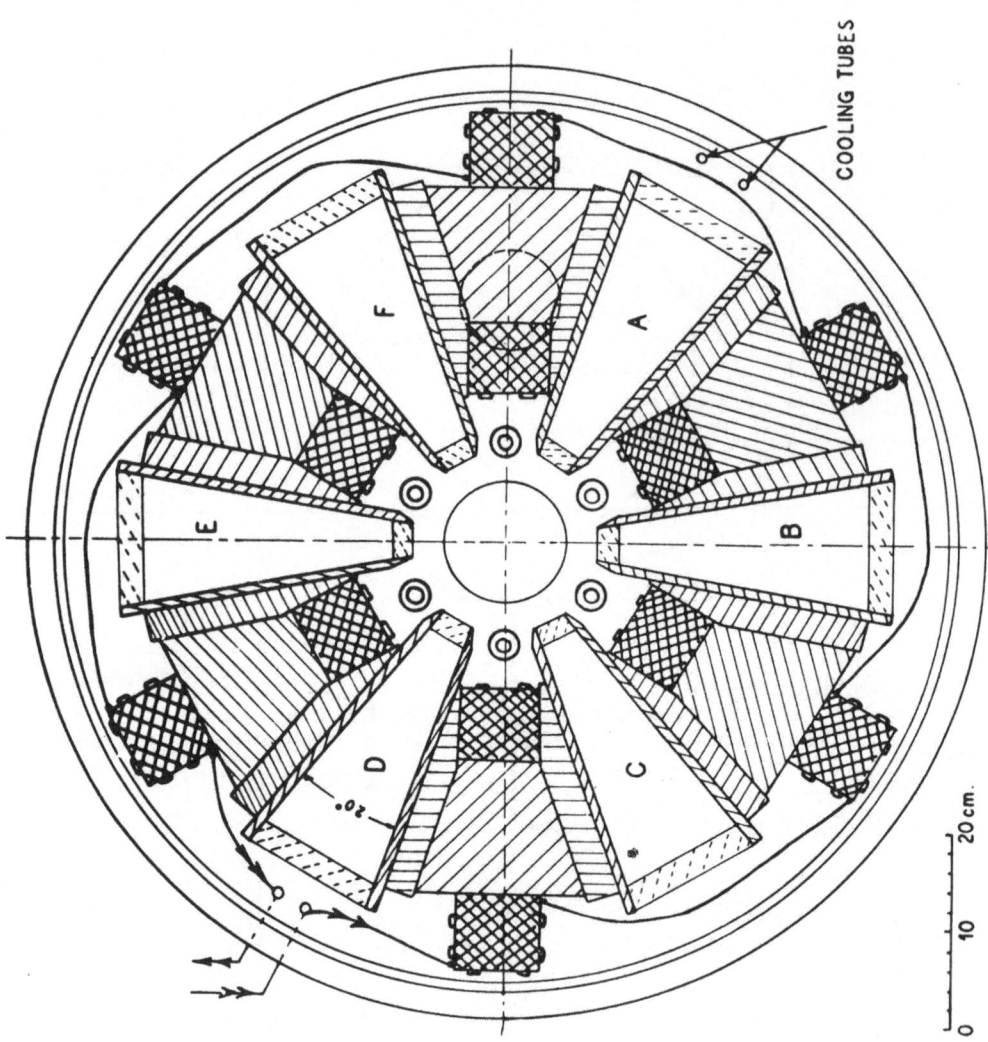

COOLING TUBES

20 cm.

10

0

11.11

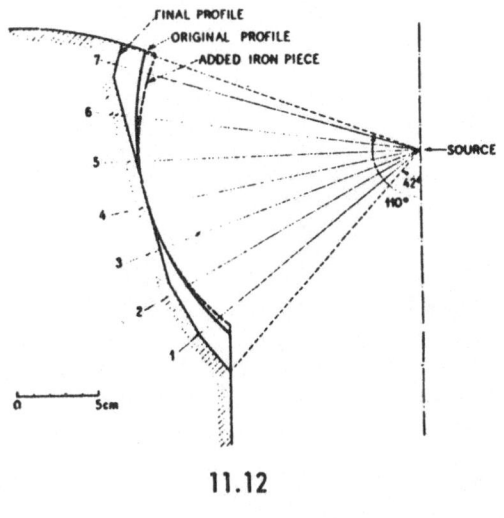

11.12

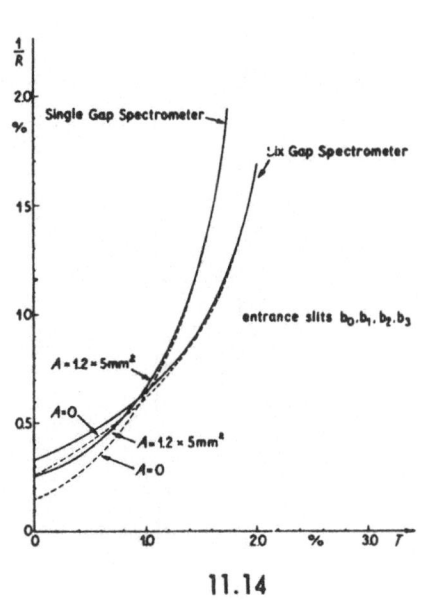

11.14

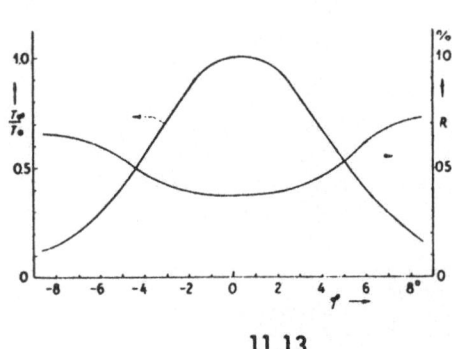

11.13

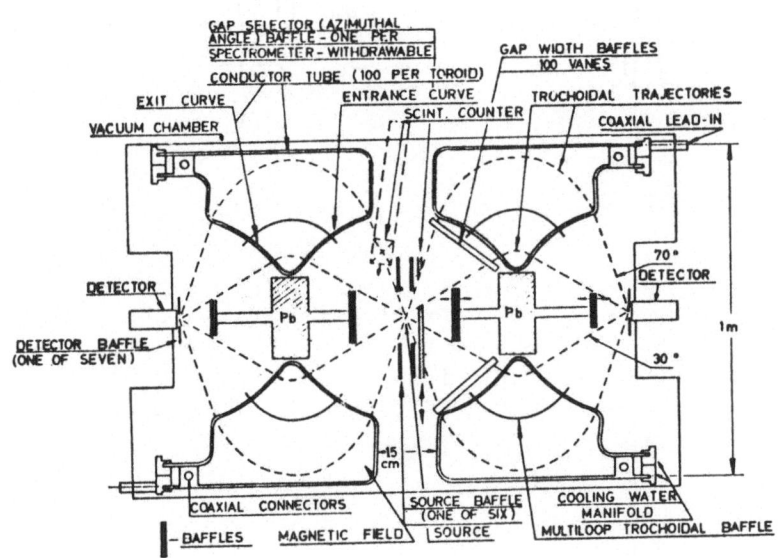

11.15

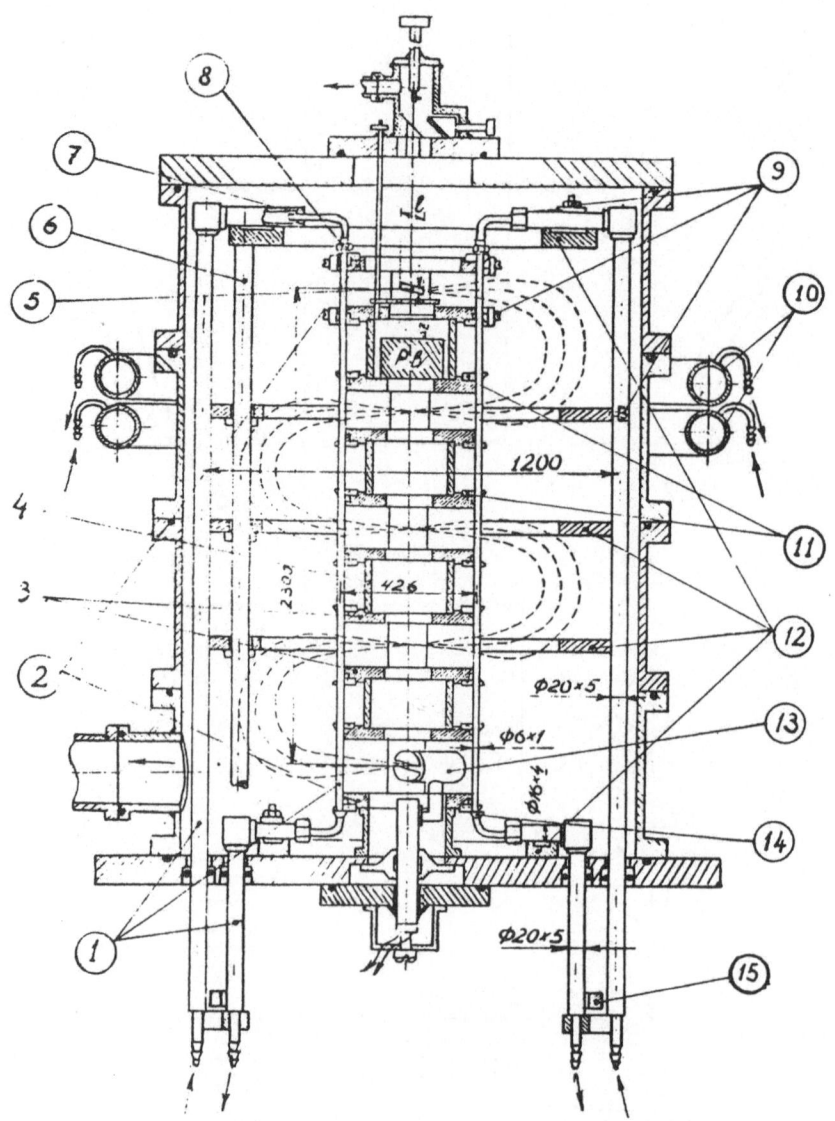

11.16

12. TROCHOIDAL SPECTROMETERS

Thibaud (1) was the first to use a magnetic field strongly decreasing radially, to separate electrons from positrons and detect them at oposite ends of the magnet. Particles were desribing trochoidal orbits. The apparatus was not a spectrometer since there was no energy discrimination, but rather a device to separate particles of opposite sign and bring some of them to detector.

Thibaud's work in Lyon was continued by Lafoucrière (2-6), who studied first the focusing propertien of r^{-n} fields in a symmetry (median) plane perpendicalar to the axis of cyllindrical symmetry. He found that r^{-1} field produces most satisfactory focusing in the median plane. Monokinetic particles emitted in any direction from a source placed on a circle, all return to another point on the same circle. For $n \leqslant 1$ this is not possible, as illustrated in Fig.1. The properties of skew rays and designs of spectrometer were then studied by a number of Lafoucrieres collaborators: Bastard (7-10) and Riche (11), who were followed by Lucenet (12-14), who built a small iron-core trochoidal spectrometer and Mugnier (15-21) who studied the design of a high performance iron-free spectrometer.

Lee-Whiting (22) made a detailed theoretical study of the focusing properties of r^{-1} field and design parameters of a high transmission - low resolution trochoidal spectrometer. He also considered the calculations of an iron-core and an iron-free magnet for that spectrometer.

While Lyon group considered only single-loop spectrometer, Hofmann (23-25) studied properties of multiple-loop spectrometer and then a six-loop trochoidal iron-core spectrometer was designed and built in Zurich by Balzer, Bharucha, Heinrich and Hofmann (26-29).

In the following, we shall first derive focusing properties in median plane, then summarize Lee-Whiting s results and give brief reviews of recent work in Lyon and Zurich.

Malmfors used trochoidal geometry to devise a time of flight spectrometer (30-32). Since this type of spectrometer is not considered in this book, reader is reffered to Siegbahns review (33) and original papers.

Motion in median plane. - The source is placed at the circle having radius equal to r_o which shall be called source circle. The coordinate of the source are $r = r_o$ $\theta = 0$. The direction of electron motion is described by the angle Ψ which tangent to the circle having the radius equal to r makes with velocity vector (or, more generally, projection of velocity to the median plane). (Fig. 12.2). The initial value of Ψ is Ψ_o. The components of the velocity vector are then

$$r \, \dot\theta = v \cos\Psi \qquad\qquad (12.1a)$$

$$\dot r = v \sin \psi \qquad\qquad (12.1b)$$

The field at the source circle is $B_z(r_o, o) = B_o$ and at any other values of r it is given by

$$B_z (r, o) = B = B_o r_o / r \qquad\qquad (12.2)$$

In order to obtain a characteristic parameter it is necessary to introduce B^o, the intensity of a uniform field in which the particle of momentum p would move in a circle of radius r_o (the circle of course being perpendicular to the plane of B^o)

$$B^o = p \, (e \, r_o)^{-1} \qquad\qquad (12.3)$$

Then a parameter k can be defined as

$$k = B_o / B^o = e \, r_o B_o / p = e \, Br/p = \frac{r}{\rho} \qquad\qquad (12.4)$$

where ρ is the radius of curvature which the electron would have for field strength equal to B. It is obvious that only values of $k > 1$ are of interest, since electron could not return to source circle, once it has $r > r_o$. It will be seen below that k is the principal parameter for the description of spectrometer characteristics.

The electron trajectory can be found starting from equations of motion (4.12)

$$(m/e)(\ddot r - r \, \dot\theta^2) = - \dot\theta \, r \, B_z \qquad\qquad (12.5)$$

$$(m/e) \frac{d}{dt} (r^2 \dot{\theta}) = r \dot{r} B_z \qquad (12.6)$$

and puting B_z given by (2), they become

$$(m/e) (\ddot{r} - r \dot{\theta}^2) = - \dot{\theta} B_o r_o \qquad (12.7)$$

$$\frac{d}{dt} (r^2 \dot{\theta} - (e/m)r_o B_o r) = 0 \qquad (12.8)$$

Equation (8) can be integrated to

$$r^2 \dot{\theta} - (e/m)r_o B_o r = C \qquad (12.9)$$

The integration constant C is obtained from the initial conditions $r = r_o$, $r\dot{\theta} = v \cos \Psi_o$, so that (9) becomes

$$r^2 \dot{\theta} - (e/m)r_o B_o r = r_o v \cos \psi_o - (e/m)r_o^2 B \qquad (12.10)$$

Expressing mv as function of k, with the help of (4), equation (10) becomes

$$(r \cos \Psi)/k - r = (r_o \cos \psi_o)/k - r_o$$

giving

$$r = r_o (k - \cos \Psi_o)/(k - \cos \psi) \qquad (12.11)$$

Relation (11) shows that particle returns to the same radius r after ψ changes by 2π , 4π , etc.

Parametric description of θ - coordinate can be found by taking the ratio of (1a) to (1b)

$$r \, d\theta / dr = \cos \Psi /\sin \Psi \qquad (12.12)$$

From (11) one finds that dr/r is given by

$$dr/r = [\sin \Psi/(k - \cos \Psi)] \, d\Psi \qquad (12.13)$$

From (12) and (13) it follows that

$$d\theta = [\cos \Psi/(k - \cos \Psi)] \, d\Psi \qquad (12.14)$$

giving the integral

$$\theta = \int_{\Psi_0}^{\Psi} \left[\cos \Psi /(k - \cos \Psi) \right] \ d\Psi \tag{12.15}$$

The solution is

$$\theta = - \Psi_0 + \Psi + 2k(k^2+1)^{-1/2} \left[\tan^{-1}(k+1)^{1/2}(k-1)^{1/2}\tan \Psi_0/2 - \right.$$

$$\left. - \tan^{-1}(k+1)^{1/2}(k-1)^{1/2} \tan \Psi /2 \right] \tag{12.16}$$

Trajectories described by (11) and (16) have the form of trochoids shown in Fig.12.2.

Electron starting from the source returns to the source-circle after an angle θ_f given by

$$\theta_f = \int_0^{2\pi} \left[\cos \Psi /(k - \cos \Psi) \right] \ d\Psi$$

$$= 2\pi \left[k(k^2-1)^{-1/2} - 1 \right] \tag{12.17}$$

Equation shows that θ_f does not depend on emission angle Ψ_0, as it was first noticed by Lafoucriere (4). There is therefore a perfect focusing in median plane for particles emitted with any walue of Ψ_0. Furthermore, since θ_f neither depends on r_0, a line source will also have a perfect image, which is obtained by rotating the source through an angle θ_f about z-axis.

It can also be seen from (17) that focusing angle θ_f increases as k decreases from k > 1. At the limiting value of k → 1, focusing angle tends to infinity. The numerical relation between k and θ_f is shown on Fig.12.3. The focusing angle is less than 2π for k ⩾ 1,154.

Dispersion. - It can be seen from relation (4) that if r_0 and B_0 are the same and source emit electrons with somewhat increased momentum p+ Δp, the value of k would decrease to k - Δk. Focusing angle would then be smaller by an amount $\Delta \theta_f$. The electrons having a momentum difference Δk would reach the source - circle at two points, separated by a distance Δs equal to

$$\Delta s = r_0 \ \Delta \theta_f$$

$$= - r_0 2\pi \ (k^2 - 1)^{-3/2} \ \Delta k \tag{12.18}$$

Dispersion D_r can be defined as

$$D_r \equiv (\Delta s/r_o)/(\Delta p/p) \tag{12.19}$$

From (18) and relation between Δk, Δp and p obtained by differentiating (4), the dispersion becomes

$$D_r = 2\pi \; k.(k^2 - 1)^{-3/2} \tag{12.20}$$

The dispersion also increases when k decreases and $D_r \to \infty$ when $k \to 1$. Large focusing angles θ_f give large dispersion. It can be seen from Fig.12.3 that dispersion is greater than 25 for $\theta_f = 360^o$ which means that in principle a high dispersion could be obtained even for $\theta_f < 360.^o$

The relation (20) is not valid for $\Psi_o = 0$ and π, because the particles are then emitted in the direction of dispersion. For tangetial emission the dispersive action of the field is similar to that of a lens. It is discussed by Lee-Whiting (22).

It should be noted that (20) gives the dispersion for one loop. For two loops the dispersion is twice larger, and in general for n loops the dispersion is nD_r.

Optimal radii.- A magnitude of great practical interest is the radial extension of the magnet over which the r^{-1} field has to be fitted. A single trochoide spreads over a radial region $(r_+ - r_-)$, where optimum radius r_+ is obtained for $\Psi_o = 0$ and minimum radius r_- for $\Psi_o = \pi$. It follows then from (11) that

$$r_+ = v_o (k - \cos \; \Psi_o)/(k - 1) \tag{12.21}$$

$$r_- = v_o (k - \cos \; \Psi_o)/(k + 1) \tag{12.22}$$

and

$$v_+/v_- = (k + 1)/(k - 1) \tag{12.23}$$

Relation (23) shows that the relative radial extension of a single trochoide increases when k decreases aproaching $k = 1$. This shows what kind of a price has to be payed for high dispersion in a single loop instrument. High dispersion requires values of k close to 1, and then not only θ_f increases, but also r_+/r_-, requiring relatively larger radial region of magnetic field to follow r^{-1} law. The ratio of r_+/r_- is also shown on Fig.12.3, which is taken from Lee-Whitings paper (22).

If all the trochoides, passing a given entrance opening, are taken into account, the bundle will have a maximum radial extent r_{max} and minimum one r_{min}, depending on opening angles Ψ_1 and Ψ_2. In the case of radial emission between angles $\pi - \Psi$ and Ψ, the r_{max} will be given by r_+ of the trajectory starting at $\Psi_0 = \pi - \Psi$ and r_{min} by r_- of the trajectory passing at $\Psi_0 = \Psi$. Their ratio is then

$$r_{max}/r_{min} = \left[(k+1)/(k-1) \right] \quad \left[(k + \cos \Psi_0)/(k - \cos \Psi_0) \right] \quad (12.24)$$

<u>Lee-Whiting's calculations</u>. - The simple analitical treatment of various characteristics of trochoidal spectrometer is strictly limited to median plane. The skew orbits can only be properly studied by numerical integration of equations of motion. Lee-Whiting proceeds in two steps, first considering only those skew rays which are emitted at a very small angle ω with respect to median plane and looking for values of k - parameter which give good focusing. In the next step large emission angles ω are considered for particular values of k.

In the first step it is supposed that a point source emits rays at a very small angle ω and equations of motion are numerically solved to find the position of the point of intersection of these rays with the right cylinder erected on the source circle, called focal cylinder. The coordinates of the point of intersection are

$$r = r_0 \qquad \theta = \theta_f + \Delta \qquad z = z_f \qquad (12.25)$$

where Δ is the important aber ration contributing to the resolution, while z_f only contributes to the height of detector. Lee-Whiting assumes that Δ and z can be represented in the form

$$\Delta = A \omega^2 + C \omega^4 + \cdots \qquad (12.26)$$

$$z_f/r_0 = B \omega + E \omega^3 + \cdots \qquad (12.27)$$

the parities being determined by the symmetry of the system.

Lee-Whiting explored the region of $1,2 \leq$ k $<$ 2 and
$-\pi /2 \leq \Psi_o$ $< \pi/2$ looking for values of k and Ψ for which A or B is zero.
In the case that A and B vanish simultaneously, the remaining radial aberrations would be of the fourth order, for a point source approximation. The
results are shown on Fig.12.4 where contours of A = O and B = O are plotted
as functions of k and Ψ_o. Only those points are acceptable where both A
and B vanish, since one of them can be very large near zero contours of
another.

Lee-Whiting finds that $\Psi_o = 0$, k = 1,291 where A and B coun-
tours are crossing, does not represent an acceptable choice because the dispersion is low. On the other hand k = 1,25 and $\Psi_o > 0.4$ seem promising.

The above results were obtained in the point source approximation.
A rectangular source placed with longer side paralel to the axis, and smaller
side approximatelly along the source circle would contribute its own width to
the image width and there will be a cross term of second order (ωz_o) between
the height of the source and the emission angle. Similar term is accidentally
cancelled in $\pi\sqrt{2}$ spectrometers. The existence of this term is in the opininn
of Lee-Whiting one of the major defects of this type of spectrometers.

In the second part of his paper Lee-Whiting has studied a spectrometer with k = 1,25. Orbit were computed for $\omega = 0 - 0,4$ rad and
$\Psi_o = 0,8 - 2,4$ rad. He finds that the largest permitted axial emission angle
ω, for a given resolution depends on Ψ_o. At R = 0,01% and for point
source, largest ω can be used with $\Psi_o \approx \pi /2$, but simultaneously image
height z_f would have maximum value. The transmission is then, for point
source, T = 2,3% for R = 0,01% and T = 6,0% for R = 0,1%, which is about
an order of magnitude better than the theoretical point source performance of
$\pi\sqrt{2}$ spectrometer.

The difficulties begin when finite source size is taken into account,
because the term proportional to ωz_o puts a severe limit on the acceptable
source height z_o. Lee-Whiting compared the luminosity of a trochoidal spectrometer with that of a $\pi\sqrt{2}$ spectrometer, assuming that their sizes would be
equal if r_{max} of the former is equal to r_o of the latter. If $r_{max} = 100$ cm

the r_o of trochoidal spectrometer would be only $r_o = 18,2$ cm, reducing overall source size. Lee-Whiting finds that for the same size and resolution, the luminosity of $\pi\sqrt{2}$ spectrometer is five times better than that of trochoidal spectrometer.

For $k = 1,25$ the ratio of maximum to minimum value of r is $r_{max}/r_{min} > 20$, depending on the transmission. The r^{-1} field would have to be fitted over relatively wery large part of the median plane. This represent a serious design problem. Lee-Whiting discusses also the magnet design. We shall refer to his proposalo further below, when discussing the work on magnet design which has been done in Lyon.

<u>Mugnier's calculations.</u> - Mugnier starts from most important design parameters, r_{max} and r_{min}, asking how transmission and dispersion would depend on mean emission angle Ψ_m, for a given r_{max}/r_{min} The dependance can be very roughly estimated by considering the focusing in the median plane.

Since axial aperture has to be small, transmission depends much on the radial aperture 2ε. For a given r_{max}/r_{min} ratio, the magnitude of 2ε varies with the mean radial emission angle Ψ_m. The relation between r_{max}/r_{min}, Ψ_m and ε can be obtained from (21) and (22), and has the following from

$$r_{max}/r_{min} = (k+1)/(k-1). \left[k+\cos(\Psi_m - \varepsilon) \right] / \left[k+\cos(\Psi_m + \varepsilon) \right] \quad (12.28)$$

The ratio r_{max}/r_{min} vary between $(k+1)/(k-1)$ and $(k+1)^2/(k-1)^2$. The largest ε is obtained for $\Psi_m = \pi$, somewhat smaller for $\Psi_m = 0$ and smallest for $\Psi_m = \pm \pi/2$. The transmission should be larger for tangential emission.

One kind of dependence of dispersion on mean emission angle Ψ_m, for a given r_{max}/r_{min}, can be obtained from the fact that source-circle radius r_o varies with Ψ_m. Since dispersion is proportional to r_o, larger possible value of r_o within a given r_{max}/r_{min} alows larger source to be used. It turns out that the largest r_o is possible with $\Psi_m = 0$, smaller with $\Psi_m = \pi/2$ and smallest for $\Psi_m = \pi$. Mugnier discards the $\Psi_m = 0$ solution because

it would be difficult to discriminate ghost peaks produced by multiple loop trajectories. The ratio between the values of r_o for $\Psi_m = \pi/2$ and π is given
by

$$(r_o)_\pi / (r_o)_{\pi/2} = (2k/k+1)\cdot \left[1+(k+1)/(k-1)\cdot(r_{min}/r_{max})\right]^{-1} \quad (12.29)$$

This relation, favouring radial emission, shows that for $k = 1,3$
and $r_{max}/r_{min} = 8$, the radius of source-circle would be about twice larger
for $\Psi_m = \pi/2$, than for $\Psi_m = \pi$.

Mugnier then proceeds to computation of trajectories for emission
angles from $\pi/2$ to $3\pi/2$. One should note that Lee-Whiting covered the
emission angles in forward direction, towards the detector, from $\pi/2$ to
$-\pi/2$, and Mugnier completes the calculations with backward emission angles.

For a point source, Mugnier finds that the spherical aberrations
are relatively smallest for

$k = 1,288$ and $2,25$ for tangential emission

$k = 1,288$ and $2,15$ for radial emission

and that at the same resolutions, the tangential emission ($\Psi_m = \pi$) has about
5-10 times better transmission.

The results for sources of finite size and $r_o = 100$ cm are given in
the table below.

TABLE 12.I

m	k	Source		Detector slit		T %	L cm^2	R = dk/k
		width, cm	height, cm	width, cm	height, cm			
π	2,25	0,2	5	0,1	5	4,7	0,047	$1,2 \times 10^{-3}$
π	2,25	0,1	5	0,05	5	3,2	0,016	$5,8 \times 10^{-4}$
$\pi/2$	2,15	0,2	3	0,1	5	0,63	0,04	1×10^{-3}
π	1,288	0,2	7	0,1	5	4,8	0,068	$1,3 \times 10^{-4}$
$\pi/2$	1,288	0,2	2	0,1	5	0,22	0,0005	$1,3 \times 10^{-4}$

The tangential emission appears to be superior, with high transmissions. The source heights are greater than expected and the reason might be

that the coefficient of the cross term ωz_o becomes very small for the parameters used.

Magnet design for $k < 1,3$.- The ratio of maximum to minimum orbit distance from the axis increases rapidly when lower values of k are approached, as illustrated by the following two examples

k	r_{max}/r_{min}
2,15	2,7
1,29	8

Magnet design problems can be illustrated by following data, taken from a more extensive table, prepared by Lucenet (14). They represent the parameters of an iron-core magnet, for measurement of electron energies up to 2,5 MeV, having source-circle radius $r_o = 20$ cm.

TABLE 12.II

k	Pole pieces		Focusing angle θ_f	Volume of vacuum chamber, lit.
	diameter, m	weight, tons		
1,3	3	70	203^0	1400
1,5	2	18	123^0	320
1,7	1,5	8	85^0	150
2	1,2	3,6	55^0	40

By increasing the source circle radius from $r_o = 20$ cm to $r_o = 30$ (k = 1,3) the weight of the magnet becomes 200 tons! For comparison, let us mention that $r_o = 50$ cm iron-core $\pi\sqrt{2}$ magnet of Nobel Institute weghts 4 tons. The size of the magnet appears to be out of proportions with any so far built for measurement of low energy electrons. That might not be the main difficulty, because it is hard to put a price for a high performance. The development of nuclear physics has shown that the size and funds are not the problem, if new research possibilities are offered. In the light of experience gained up to now on development of iron-core spectrometers, more serious difficulties might be caused by relatively very large radial extension over which the r^{-1} field has to be fitted. In $\pi\sqrt{2}$ spectrometers the field has to be fitted over

r_{max}/r_{min} < 1.5, and still there is hardly any iron-core spectrometer giving the desired field with high accuracy over the whole energy interval.

The design of pole pieces was discussed by Lee-Whiting (22) and he finds that their shape is conical, except for a small region near the axis.

Air-cored coils offer the advantage that once their geometry was computed, no serious surprises are to be expected and the field would be such as calculated.

Lee-Whiting also proposed the coil geometry for the r^{-1} field. It is based on two half cones, separated and facing each other by their tips. The current in the coils has to fall as r^{-1}. Two correcting pairs of coils have to be added, one at small radius to correct for distance between coils, and another at large radius, to correct for finite height of half cones.

Mugnier developed further Lee-Whitings proposed geometry (21). He prefers to cut the cone tips where currents become too large, and correct for it with a pair of coils. He has calculated these colls, as well as large cones for corection of halfcones final height. He also has a proposal, in principle very simple, to produce the current density proportional to r^{-1}. It the half--cone is composed from a number of coils each with a different radius and each independently connected with the same voltage supplies, the current would fall as r^{-1}, because the resistance of coils increases with r. Practical realisation would probably not be easy.

Six loops trochoidal spectrometer.-   The design of the six loops trochoidal spectrometer built in Zurich (24-29) was based on theretical calculations made by Hofman (23). He uses series expansion approach, the small quantities being axial emission angle ω_o and the height of the source measured in spherical coordinates by the angle $\alpha_o = z_f/r_o$. The three spherical coordinates of the particle have then the form

$$r(\Psi) = R_o(\Psi) + R_2(\Psi)\,\omega_o^2 + R_{11}(\Psi)\,\omega_o\,\alpha_o + R_{02}(\Psi)\,\alpha_o^2 + \ldots \quad (12.30)$$

$$\theta(\Psi) = \theta_o(\Psi) + \theta_{20}(\Psi)\,\omega_o^2 + \theta_{11}(\Psi)\,\omega_o\,\alpha_o + \theta_{02}(\Psi)\,\alpha_o^2 + \ldots \quad (12.31)$$

$$\alpha(\Psi) = \alpha_{11}(\Psi)\,\omega_o + \alpha_{01}(\Psi)\,\alpha_o + \alpha_{30}(\Psi)\,\omega_o^3 + \ldots \quad (12.32)$$

Putting (30-32) into equations of motion he proceeds in usual way, finding first and second order solutions. He finds then that for certain values of k, which he calls "stable" ones, the first and second order aberrations disappear for a definite number m of loops, larger than one.

For three "stable" values of k Hofmann calculated the performance parameter by numerical integration of equations of motion. This was necessary because the series approach is valid only for small values of axial emission angle ω_o. His results for r_{max} = 1 m are given below.

TABLE 12.III

k	m	$m\theta_f$	D, rad	pole pieces opening β	$\Delta p/p = 10^{-3}$		$\Delta p/p = 10^{-4}$	
					T%	L, cm²	T%	L, cm²
3,22	2	37°	1,4	10	2,1*	2,8x10⁻²	2	2,4x10⁻³
				20	3,0	7,2x10⁻²	2	2,7x10⁻³
1,33	2	373°	25	10	1,1*	2,3x10⁻²	1,1	2,3x10⁻³
				20	2,1*	8,0x10⁻²	2,0	6,6x10⁻³
1,31	3	549°	41	10	1,3*	3,2x10⁻²	1,3	3,2x10⁻³
				20	2,6	1,1x10⁻²	2,5	1,5x10⁻²

(Asteric indicates that transmission is limited by angle β). Although most of the T values are limited by small opening of pole pieces, both T and L are relatively large.

The spectrometer built in Zurich has k = 3,22 and the number of loops m = 2 multiplied by n = 3, to increase the dispersion. The parameters of spectrometer are

k	number of loops, nm	total focusing angle	total dispersion n m D	r_{max}/r_{min}
3,22	6	112°	4,2	2,6

The spectrometer is of iron-core type with rather reduced dimensions, r_{max} being only 25 cm. The magnet is shown in Fig.12.5. and the cross section of multiple aperture system in Fig.12.6. The source and slits have to be very accurately adjusted. The source position is not easily reached and in present design the whole geometry difining system has to be removed from the magnet when a new source is introduced.

The actual performance of spectrometer falls below the theoretical predictions, as shown on Fig.12.7. At high resolutions the transmision is several times smaller than expected, and the source size contribution is more important than predicted. This is caused in the opininion of authors by "the residual errors existing in magnetic field".

The performances, although below predictious, are still very good for such a small spectrometer. The source is very narow (< 1 mm) but 3 cm high, even for R ≈ 0,045%, which contributes to good L – R characteristics.

References for chapter 12.

1. J.Thibaud, C.R.Acad. Sc., 197, 447 (1933).

2. J.Lafoucrière, C.R.Acad.Sc., 229, 823 (1949).

3. "   "   C.R.Acad.Sc., 229, 1005 (1949).

4. "   "   C.R.Acad.Sc., 231, 137 (1950).

5. "   "   Ann.Phys. 6, 610 (1951)

6. A.Moussa and J.Lafoucriere, C.R.Acad.Sc.233, 139 (1951)

7. C.Bastard and J.Lafoucrière, J.Phys.Rad., 19, 674 (1958)

8. C.Bastard, J.Lafoucrière and R.Margrita, J.Phys.Rad., 20, 736 (1959)

9. C.Bastard, and J.Lafoucrière, J.Phys. Rad., 21, 112 (1960)

10. C.Bastard, C.R.Acad. Sc., 248, 3295 (1959)

11. J.Riche, Nucl.Instr.Meth. 32, 157 (1965)

12. G.Lucenet, Nucl.Instr. Meth. 24, 51 (1963)

13. G.Lucenet,Nucl.Instr.Meth. 45, 250 (1966)

14. G.Lucenet, These Doc.Ing. Lyon 1966

15. D.Mugnier and J.Lafoucrière, C.R.Acad.Sc. 259, 1098 (1964)

16. D.Mugnier and J.Lafoucrière, Nucl.Instr.Meth. 45, 45 (1966)

17. D.Mugnier, R.Gayet and J.Lafoucrière, C.R.Acad.Sc. 262, 75 (1966)

18. D.Mugnier, P.Argout and J.Lafoucrière, Nucl.Instr.Meth. 41 277 (1966)

19. D.Mugnier, P.Beuzit and J.Lafoucrière, Nucl.Instr.Meth. 50, 77 (1967)

20. D.Mugnier, J.Josepf and J.Lafoucrière, C.R.Acad.Sc. 263, 576 (1966)

21. D.Mugnier, These Doct. Sc., Lyon 1968.

22. G.E.Lee-Whiting, Canad. J.Phys., 41, 496 (1963).

23. A.Hofman, Nucl.Instr.Meth 40, 13 (1966)

24. F.Heinrich and A.Hofman, Helv.Phys.Acta 35, 322 (1962)

25. A.Hofman, Helv. Phys. Acta 36, 814 (1963)

26. R.Balzer, D.Bharucha, F.Heinrich and H.Hofman, Helv Phys. Acta 37, 602 (1964)

27. R.Balzer, D.Bharucha, F.Heinrich and A.Hofmann, Helv. Phys. Acta 37, 603 (1964)

28. R.Balzer, D.Bharucha, F.Heinrich and A.Hofmann, Helv.Phys. Acta 40, 197 (1967).

29. R. Balzer, D. Bharucha, F. Heinrich and A. Hofmann, Nucl. Instr. Meth. <u>57</u>, 277 (1967)

30. K. G. Malmfors, Nucl. Instr. Meth. <u>1</u>, 251 (1957)

31. K. G. Malmfors, Ark. f. Fysik <u>13</u>, 237 (1958)

32. K. G. Malmfors and A. Nilsson, Ark. f. Fysik, 247 (1958)

33. K. Siegbahn (editor) α, β and γ -Ray Spectroscopy, North-Holland (1965), p. 140.

Text to Figures. Ch.12.

Fig.12.1. Focusing in the median plane, for fields r^{-n} when $n \gtrless 0$. Electrons emitted at $\psi_0 = 0$ and π do not return to the source circle at the same point. One can intuitively see that the return point should coincide for $n = 1$.

Fig.12.2. Parameters defining the trochoidal orbits.

Fig.12.3. Dispersion D_r, focusing angle θ_f and ratio of limiting circles radii r_+/r_- shown as functions of parameter k.

Fig.12.4. Contours $A = 0$ and $B = 0$ shown as functions of emission angle ψ_0 and parameter k. Curves labeled A and B represent the foci of points $A = 0$ and $B = 0$ respectively.

Fig.12.5. The magnet design for six-loop spectrometer.

Fig.12.6. The geometry of Zurich six-loop spectrometer.

Fig.12.7. a) Resolution in function of the source width.

b) Transmission in function of the resolution.

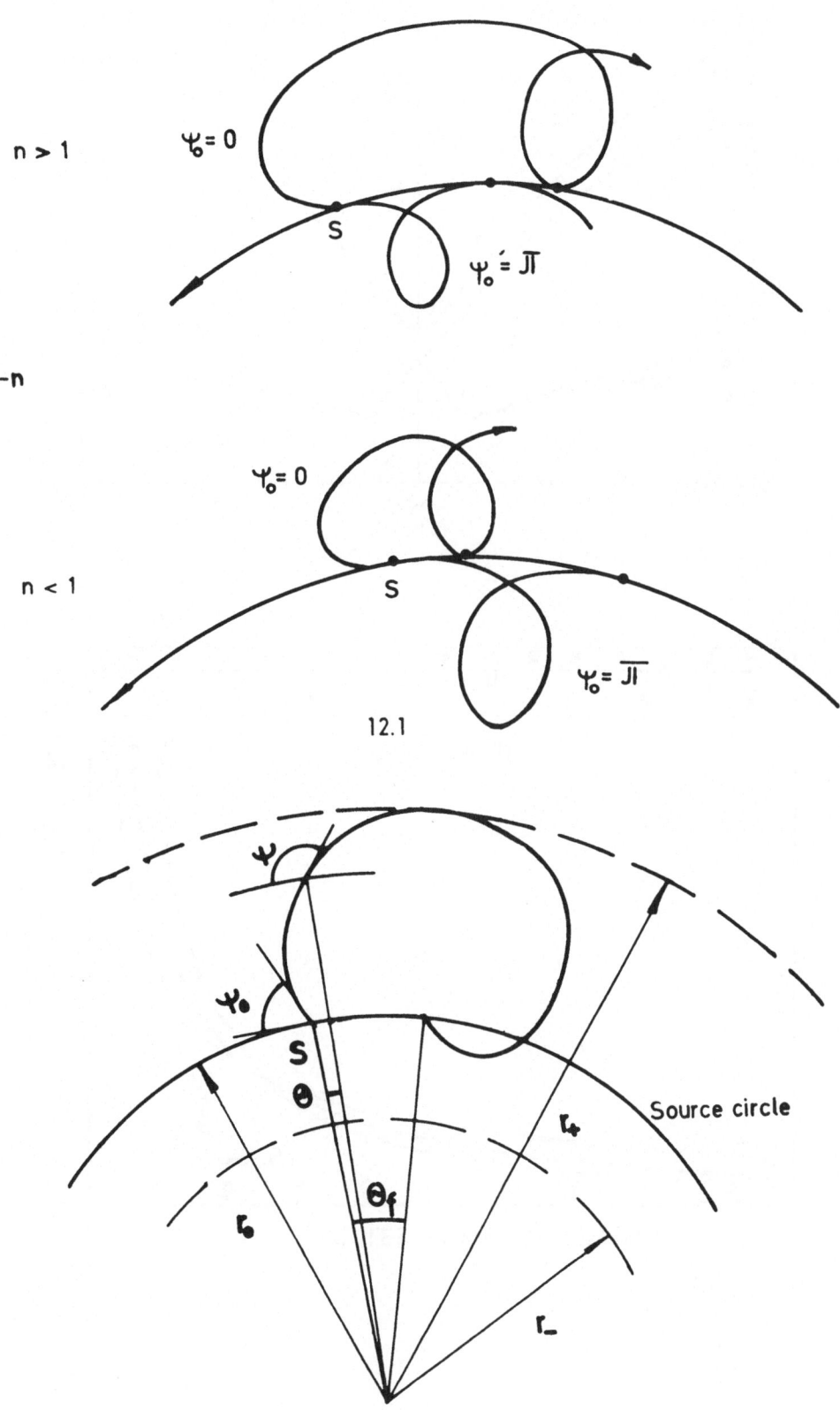

$n > 1$

$\psi_0 = 0$

$\psi_0' = \pi$

$B \sim r^{-n}$

$n < 1$

$\psi_0 = 0$

S

$\psi_0 = \overline{\pi}$

12.1

ψ

ψ_0

S

θ

r_+

Source circle

r_0

θ_f

r_-

12.2

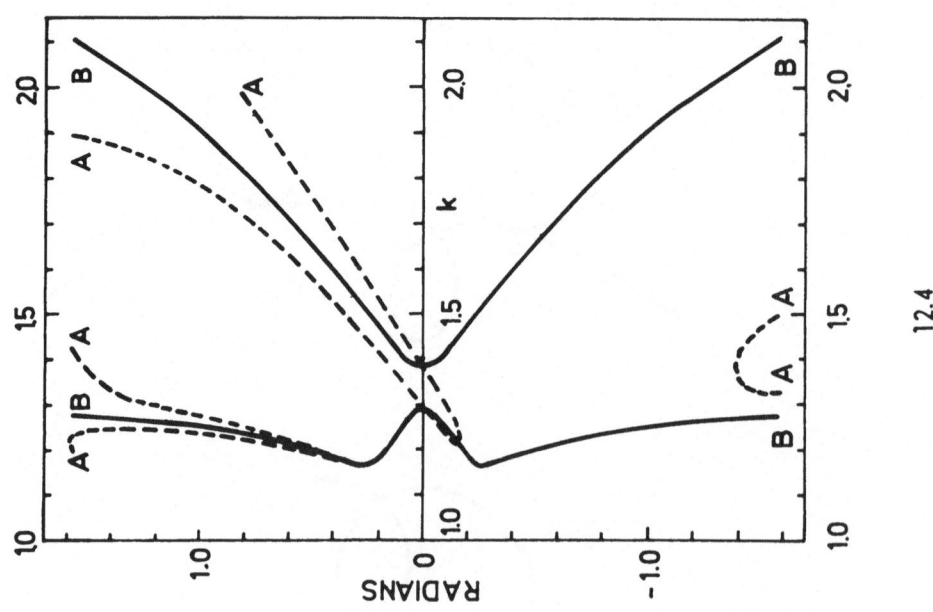

12.4

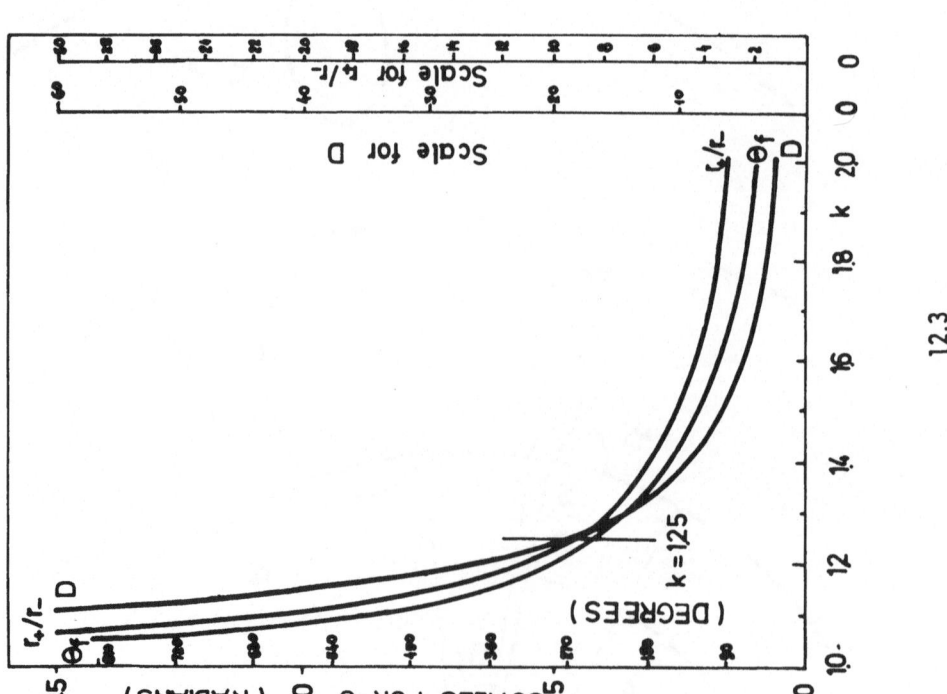

12.3

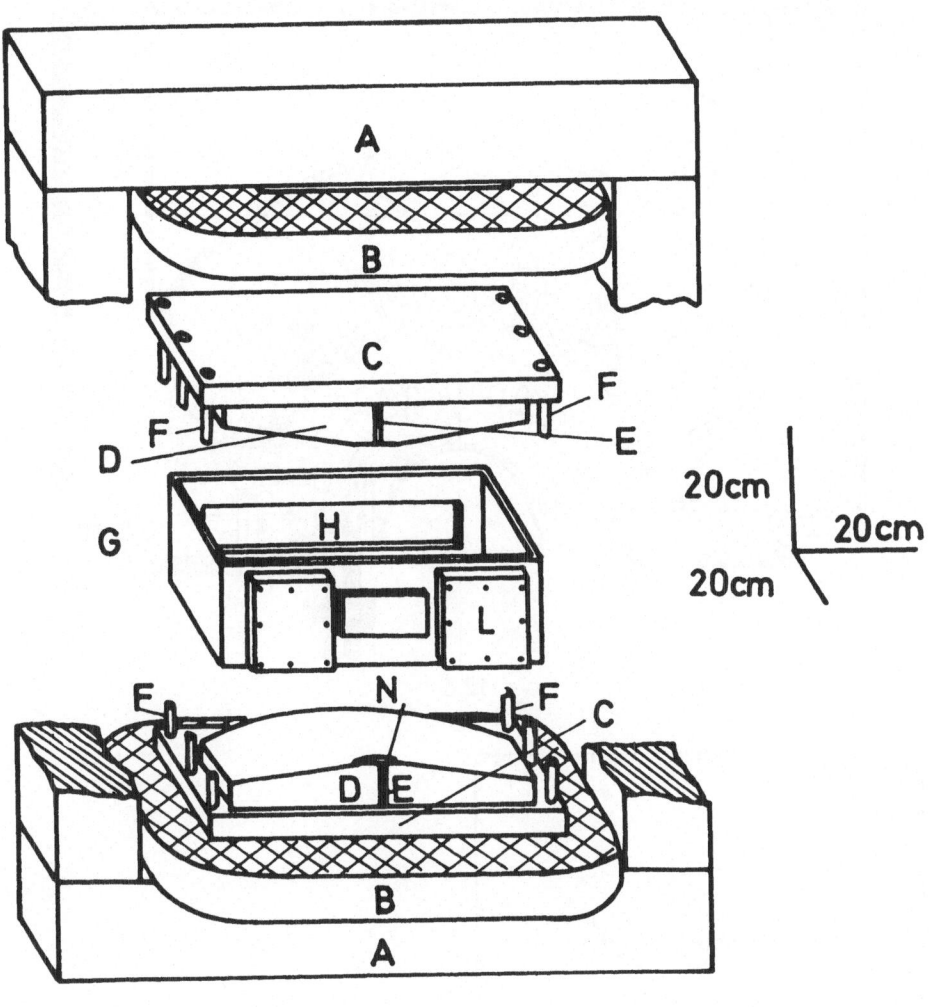

20cm

20cm

20cm

12.5

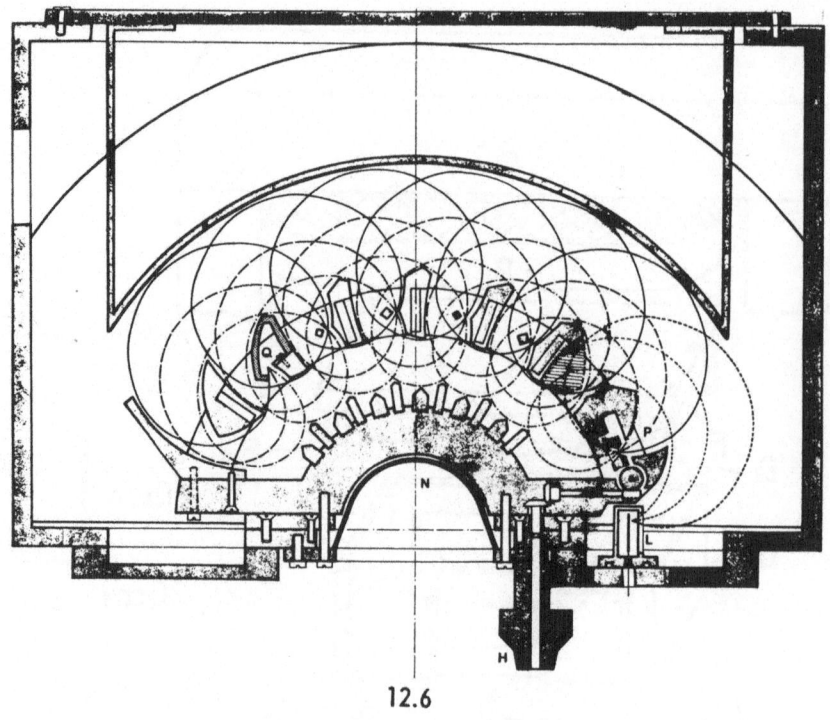

12.6

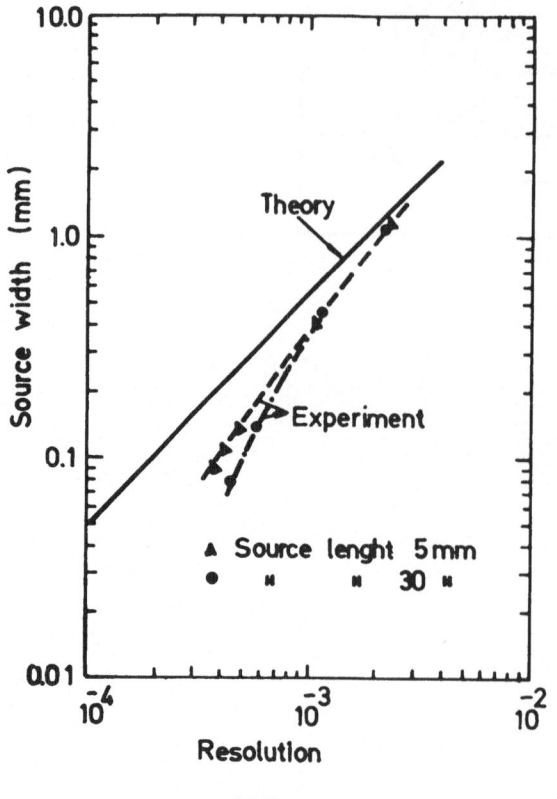

12.7. a

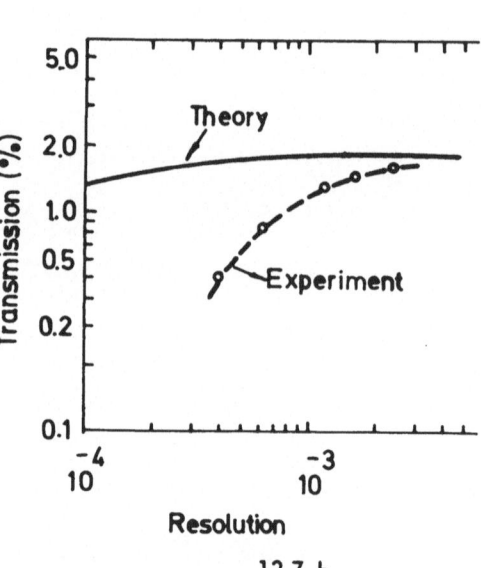

12.7. b

L E N S E S

The lens spectrometers, especially short lenses, are easiest to construct. Many different field forme were used and there are more lenses reported in literature than ony other type of betá spectrometers.

A complete analytical treatment of focusing properties is possible only in the case of uniform field. The expressions thus derived can serve as a guide for understanding the work of lenses in general. For that reason we shall start with solenoidal field and derive the focusing properties, following the treatment of Persico (1) and du Mond (2).

Another typical approach represents the paraxial approximation in thin lens which yields the formulae analogous to geometrical optics. The paraxial beams are used in electron microscopes, while the almost always present need for good transmission of beta spectrometers requires larger departure angles, for which the paraxial approximation is not valid any more. We shall only derive the basic thin lens formulae, referring the reader for more details to extensive literature on electron microscope optics.

In most of the other cases the lens properties can be only studied by tracing the rays. While earlier the beam geometry was investigated with the help of an electron gun and fluorescent screen, or a longitudinally placed photographic plate, the recent development of computers allows rather complete theoretical studies.

13. LONG LENS

Electron trajectories. In a solenoidal spectrometer the field is homogeneous and parallel to an axis containing the source and the detector. The entrance baffle accepts electrons emitted between the angles γ_1 and γ_2 with respect to the axis. The beam possesses the cylindrical symmetry and the focusing is two-dimensional.

Taking the spectrometer axis as the z-axis of a cylindrical coordinate system, the homogeneous field reduces to

$$B_r = 0 \qquad B_\theta = 0 \qquad B_z = B \qquad (13.1)$$

and the equations of motion in cylindrical coordinates (4.13-15) become

$$\ddot{r} - r\dot{\theta} = -(e/m)\, Br\dot{\theta} \qquad (13.2)$$

$$\frac{d}{dt}(r^2\dot{\theta}) = (e/m)\, Br\dot{r} \qquad (13.3)$$

$$\ddot{z} = 0 \qquad (13.4)$$

The initial conditions for an electron leaving from a point source at the axis, with an angle γ with respect to z-axis, are

$$z_0 = 0 \quad r_0 = 0 \quad (\theta)_0 = \theta_0 \qquad (dz/dt)_0 = v_0 \cos\gamma \qquad (13.5)$$

where v_0 is the electron velocity, and θ_0 is the angle between $v_0 \sin\gamma$ and the plane taken for $\theta = 0$.

With these initial conditions the equation (4) can be immediately integrated to

$$z = v_0 t \cos\gamma \qquad (13.6)$$

The equation (3) can be written as

$$\frac{d}{dz}(r^2\theta - (e/m)(B/2)r^2) = 0 \qquad (13.7)$$

Integrating (7) once, taking into account (5) and multiplying by r^{-2}, one obtains

$$\dot{\theta} = (e/m)\, B/2 \qquad (13.8)$$

which can be integrated to

$$\theta = (e/m) \ (B/2) \ t + \theta_o \qquad\qquad (13.9)$$

Putting (8) into (2), the equation for r becomes

$$\ddot{r} + (eB/2m)^2 \ r = 0 \qquad\qquad (13.10)$$

The general solution of (10) is

$$r = C_1 \cos (eB/2m) \ t + C_2 \sin (eB/2m)t \qquad\qquad (13.11)$$

and with initial conditions (5) becomes

$$r = (2m/eB) \ v_o \sin \gamma \ \sin (eB/2m)t \qquad\qquad (13.12)$$

Introducing

$$G = (2mv_o/eB) \qquad\qquad (13.13)$$

$$\tau = (eB/2m) \ t \qquad\qquad (13.14)$$

the equations defining the trajectory can be written as

$$r = G \sin \gamma \ \sin \tau \qquad\qquad (13.15)$$

$$z = G \cos \gamma \cdot \tau \qquad\qquad (13.16)$$

These are equations of an helix (Fig.1) with diameter d and pitch L given by

$$d = G \sin \gamma \qquad\qquad (13.17)$$

$$L = \pi G \cos \gamma \qquad\qquad (13.18)$$

Equation (15) represents the projection of electron trajectory on a plane rotating with electron around z-axis. That is a convenient representation to visualise the trajectories in a lens and will be used extensively in the rest of this chapter.

Position of ring focus. Two electrons, having the same momenta and emitted under the angles γ and $\gamma + d\gamma$ (Fig.2) follow trajectories with radial coordinate r and $r + dr$ respectively. In the first half of the lens dr increases, and then in the second half decreases to become zero in plane z_{rf}

and change the sign, as shown Fig.2. The width of a beam would not be mini-mum in the focus but near z_{rf} - plane, where a ring focus is formed. The exit slit of a lens is placed in the rig focus to obtain minimum contribution of image width to the total line-width. Precise location and geometry of ring focus represents one of the most important tasks of lens design.

The two rays emitted with an angular difference $d\gamma$ form the ring focus at a value τ_F given by

$$dr/d\gamma = 0 \qquad (13.19)$$

$$dz/d\gamma = 0 \qquad (13.20)$$

which when differentiating (15) and (16) lead to

$$(\tan \ \tau_F)/ \ \tau_F = - \tan^2 \gamma \qquad (13.21)$$

The angular coordinate θ can be used instead of τ, the connec-tion between them following from (9) and (14)

$$\theta = \tau + \theta_o \qquad (13.(22)$$

Taking $\theta_o = 0$ the relations (21) and (22) give

$$(\tan \ \theta_F)/\theta_F = - \tan^2 \gamma \qquad (13.23)$$

The numerical values connecting θ_F and γ are shown in Fig.3, taken from DuMonds paper (3).

To obtain the coordinates of the ring focus (for two orbits only), it is convenient to introduce an auxiliary function $\beta \ (\gamma)$ defined by

$$\tan \ \beta \ /(\pi - \beta) = \tan^2 \gamma \qquad (13.24)$$

Then since $- \tan \beta = \tan (\pi - \beta)$, it follows from (21), (24), (16) and (15) that the coordinates of the ring focus are

$$\tau_F = (\pi - \beta) \qquad (13.25)$$

$$z_F = (\pi - \beta) \ G \cos \gamma \qquad (13.26)$$

$$r_F = G \sin\gamma \ \sin\beta \qquad (13.27)$$

Width of ring focus. The above relations defining the position of
the ring focus were deduced considering only two rays. If only one more trajectory were added, for an emission angle of $\gamma - d\gamma$, the three trajectories
would not cross the ring focus plane at the same radial distance. Due to spherical aberations, a beam with a finite angular opening has a ring focus with a
finite width.

The angular aperture can be defined by a mean angle γ and semi-
-aperture angle ϵ , so that beam is accepted between the angles $\gamma - \epsilon$ and
$\gamma + \epsilon$. Since ϵ is small, the r-coordinates of electrons emitted at $\gamma \pm \epsilon$
differ by small amount from r-coordinate of electrons emitted at the mean
angle γ, which can be denoted as r_m . The radius of trajectory r can then
be developed in Taylor series, and the development terminated at the term of
second order. At the ring focus the term of first order vanishes because $dr/d\gamma$
must be zero there, and r is given by two remaining terms

$$r = r_I + (\partial^2 r/\partial\gamma^2)\ \epsilon^2/2 \qquad\qquad (13.28)$$

The ray tracing gives more exact picture of the ring focus geometry. As can be seen on Fig.4, its upper edge is defined by the mean ray, while
outside rays cross at its lower edge.

The width of the ring focus for a point source is therefore

$$\Delta r_p \equiv r - r_I = (\partial^2 r/\partial\gamma^2)\ \epsilon^2/2 \qquad\qquad (13.29)$$

Taking twice the partial derivative of (15) keeping in mind that τ is
also function of γ , and using (21), the width of the ring focus can be put in the
form

$$\Delta r_p = (G/2)\ (\sin\beta/\sin\gamma)\ \Gamma\epsilon^2 \qquad\qquad (13.30)$$

where $\Gamma(\gamma)$ is given by

$$\Gamma(\gamma) = (\beta + \cos^2\gamma\ \tan^2\beta) \qquad\qquad (13.31)$$

When the source has final dimensions, another two contributions are
added to the width of the ring focus, one of them due to the width of the source,
and another to the difference in emission angle from different points of the source.

Both contributions can be evaluated by considering rays emitted from different points of a disk source, when the entrance slit is infinitely thin, $\epsilon \simeq 0$. Since the maximum radial distance between two points on the source is 2s (s being the radius of the disk), this width is transmitted to the image.

A point on infinitely thin entrance slit receives the rays having emission angles between $\gamma - \delta\gamma$ and $\gamma + \delta\gamma$, where $\delta\gamma$ is proportional to the source radius. The effect on the ring focus is the same as if source were a point and entrance slit had a final aperture $2\delta\gamma$. This contribution can be taken into account by supposing that the entrance angles are $\gamma \pm (\epsilon + \delta\gamma)$. The total width of the ring focus becomes then

$$\Delta r_e = 2s + (G/2)(\sin \beta / \sin \gamma) \Gamma (\epsilon + \delta\gamma)^2 \qquad (13.32)$$

Besides being proportional to the source radius s, $\delta\gamma$ depends also on the emission angle γ and the distance z_d of the entrance diapragm from the source. The exact expression for $\delta\gamma$ is obtained from relation

$$s/\delta\gamma = (\partial r/\partial\gamma)_{z=z_d}$$

$$= G \cos\gamma \, \sin\tau + G \sin\gamma \, \cos\tau \cdot \tau \tan\gamma \qquad (13.33)$$

If the entrance diaphragm is placed half-way between the source and the detector, $\tau = \pi/2$ and (33) reduces to

$$\delta\gamma = s/(G \cos\gamma)$$

$$= \pi s/L \qquad (13.34)$$

Dispersion. In the case of the electron lens the dispersion $D = p\partial r/\partial p$ measures the radial displacement dr_F of the electron trajectory in the ring focus, when the momentum is changed by dp. Since $p = eBG/2$, dispersion D can be expressed as

$$D = G (\partial r/\partial G)_F \qquad (13.35)$$

Differentiating (15) and taking into account (21), dispersion becomes

$$D = G \sin \beta / \sin \gamma$$

$$= r_F / \sin^2 \gamma \qquad (13.36)$$

Dispersion is proportional to the radius of the ring focus and favors small emission angles. An increase in dispersion is usually achieved by increasing r_F, but the emission angles are not decreased, since that would produce a decrease of transmission.

Resolution. The base width can be defined by the magnetic field change $(\Delta B)_b$ necessary to shift the ring image by an amount equal to its own width Δr_F. The relative base width is then $(\Delta B)_b/B$, which is equal to $(\Delta p)_F/p$, where Δp_F is connected with Δr_F by the definition of dispersion $D = p \cdot \Delta r_F / \Delta p_F$. The resolution is then

$$R = (1/2) (\Delta p)_F/p$$

$$= (1/2) (\Delta r_F)/D \qquad (13.37)$$

Using (30) and (36) the resolution (37) becomes for a point source

$$R = \epsilon^2 \Gamma / 4 \qquad (13.38)$$

while (32) gives for a disk source

$$R = s \cdot \sin \gamma /(G \sin \beta) + \Gamma (\epsilon + \delta \gamma)^2 / 4 \qquad (13.39)$$

Solid angle and luminosity. The acceptance solid angle Ω, defined by entrance slit, is given by

$$\Omega = 4 \pi \epsilon \sin \gamma$$

so that the transmission becomes

$$T = \frac{\Omega}{4 \pi} = \epsilon \sin \gamma \qquad (13.40)$$

The source diameter s measured in units of G may be expressed as

$$s = G \sigma \qquad (13.41)$$

The source area A is then

$$A = \pi G^2 \sigma^2 \qquad (13.42)$$

and the luminosity L becomes

$$L = AT$$
$$= G^2 \pi \sigma^2 \epsilon \sin \gamma \tag{13.43}$$
$$= G^2 \pi \lambda$$
$$\lambda = \sigma^2 \epsilon \sin \gamma \tag{13.44}$$

where λ is proportional to the luminosity for a given G. It should be noted that luminosity is proportional to the square of G, which caracterises the linear dimensions of the aparatus, as shown by (17-18). The increase of the linear dimensions by a given ratio, brings about an increase of the luminosity proportional to the square of this ratio.

Optimum source diameter.- Persico finds the optimum source diameter by asking the question, how would the resolving power vary, when the source diameter and slit width are varied while the luminosity is kept constant. A relation between R and λ may be obtained by elimination of ϵ from (44) and (39), where $\delta\gamma$ is neglected in first approximation. The relation is

$$R = \sigma \sin \gamma / \sin \beta + \Gamma \lambda^2 / (4\sigma^4 \sin^2 \gamma) \tag{13.45}$$

The optimum value may be found by differentiating R with respect to σ , and equating to zero.

$$\frac{\partial R}{\partial \sigma} = \frac{\sin \gamma}{\sin \beta} - \frac{\Gamma \lambda^2}{\sin^2 \gamma} \cdot \frac{1}{\sigma^5} = 0 \tag{13.46}$$

Taking the value of σ from this relation as σ_o, the minimum value of 'R is obtained for source diameter $2s_o$, equal to

$$2s_o = 2G \sigma_o$$
$$= 2G \cdot \frac{\sin^{1/5} \beta \ \Gamma^{1/5} \cdot \lambda^{2/5}}{\sin^{3/5} \gamma}$$
$$= 2G \cdot \Gamma \epsilon^2 \frac{\sin \beta}{\sin \gamma} \tag{13.47}$$
$$= 4 \ \Delta r_F \tag{13.48}$$

where relations (44) and (30) have been used.

The optimum source diameter contributes to the image a partial width four times larger than the width due to aberration. Persico also calculates s_o, starting from (32) in which $\delta\gamma$ is not neglected and finds that s_o increases by a small amount of the order of a few percent, which can be neglected.

Optimum emission angle. - Resolution depends on the emission angle γ. Putting the value of σ_o into (39), the resolution may be expressed as function of γ for constant λ

$$R = \frac{5}{4} \, \lambda^{2/5} \, (\, \Gamma \, \sin^2\gamma \, \sin^{-4}\beta \,)^{1/5} \tag{13.49}$$

The optimum value of γ may be found by logarithmic differentiation of (49), and equating to zero. The solution is $\gamma \simeq 40^o$. In that case the other characteristic parameters are

$$r_F = 0,544 \, G \qquad z_F = 1,63 \, G \tag{13.50}$$

$$R_o = 1.62 \quad \lambda^{2/5} \tag{13.51}$$

$$\sigma_o = 1.69 \quad \lambda^{2/5} \tag{13.52}$$

The last two relations show that the source radius expressed in units of G is equal to the resolution, when optimum conditions are used.

In the above derivation, the term $\delta\gamma$ was neglected. Persico finds that, when $\delta\gamma$ is taken into account the optimum emission angle shifts to 36°.

From (51) and (43) one finds that the optimum relation between the resolution and luminosity, when $\delta\gamma$ is neglected, becomes

$$R_o = L^{2/5} /G^{4/5}$$

Spherical aberrations. The value of helix pitch $L = \pi \, G \cos\gamma$ tends to πG for very small emission angles. The rays emitted at very small angles γ form a paraxial focus at $z_{pf} = \pi G$. Electrons emitted at· larger angles γ return to the axis at a point z_o which is smaller than z_{pf}.

The distance $z_{pf} - z_o$ between the paraxial focus and the point, at which an electron emitted at angle γ_n, returns to axis, is called longitudinal spherical aberration ΔJ. It is equal to

$$\Delta J = G\pi - G\pi \cos\gamma_n$$
$$= G\pi (1 - \cos\gamma_n) = G\pi \left(\frac{\gamma_n^2}{2} - \frac{\gamma_n^4}{4!} + \ldots\right)$$
$$= G\pi \frac{\gamma_n^2}{2} \qquad\qquad (13.55)$$

neglecting terms of higher order than two, if γ_n is sufficiently small.

Since $G\pi = z_{pf}$, the longitudinal spherical aberration may be expressed as

$$\Delta J = \frac{z_f}{2} \cdot \gamma_n^2$$
$$= C_s \cdot \gamma_n^2 \qquad\qquad (13.56)$$

where C_s is spherical aberration coefficient, which in the case of the solenoidal lens, is equal to

$$C_s = \frac{z_f}{2} \qquad\qquad (13.57)$$

The image of a point source has final dimensions, and is extended along the axis. In the plane perpendicular to axis at z_o, the radius of the image Δr is equal to

$$\Delta r = \Delta J \tan\gamma_n$$
$$\approx C_s \gamma_n^3 \qquad\qquad (13.58)$$

This important result shows that spherical aberrations contribute to the radial dimension of the image proportionaly to the cube of the aperture.

Multiple ring focus baffle. Hubert has shown (4) that it is possible to improve the resolution by having in the ring focus not just one outside baffle but several of them one after another, with decreasing radius as detector   is approached.

Hubert baffle system is illustrated in Fig.5. where ES and DS represent the entrance slit and the detector slit respectively. The simple   ring focus baffle consists of RI and RO, so placed that their edges lie on straight lines drawn from the source. Two additional outer baffles are shown RH_1 and RH_2. It can be seen from the figure that RH_1 prevents outer rays to pass ring focus at a smaller field, while RH_2 does the same to the inner rays. Only when the field increases to proper value, all the beam passes to the detector. For that reason the line rises steeply from the low energy side.

The efficiency of the multiple ring focus baffle system increases with the number of additional outer baffles. Thus Jungerman et al (5) use 13 outer ring focus baffles. The transmission can be varied by moving axially the single inner ring focus baffle RI.

Hubert baffle is not restricted to solenoidal spectrometer, but can be used for any type of the lens.

Aberration correctors. The spherical aberrations of magnetic lenses remain positive and finite for any shape of the lens field. It can be corrected however with a transwersal field. Dolmatova and Kelman (6) designed an aberration corrector using a r^{-1} field. It is essentially an iron-free toroidal coil coaxial with the lens (Fig.6). The profiles are so designed to bring all three principal rays into an infinitely thin ring focus. The mean ray does not experience the correcting field, while the outer and inner rays are "lifted" by corrector, comming back to the axis at a point further away from the source.

At the solid angle of 8,7% (effective transmission 6,5%) and the source 1 x 1 mm^2, the resolution was with the corrector 1,9% instead of 5,7% as theoretically expected from homogenous field.

Production of uniform magnetic field. The usual way to produce a uniform magnetic field is to combine a long coil with two short coils placed at the ends of the long one, serving as end correctors. The main coil is about 2-3 times longer than source-detector distance. Such arrangement can give a magnetic field uniform within $1/10^4$, over the electron trajectories (9). More about production of uniform field can be found in references (9-12).

A more rational design was made by DuMond (3), who used ellipsoidal coil. A detailed theoretical treatment of the production of uniform field with ellipsoidal coil was made earlier by Blewett(7). The coil is made so (Fig.7) that each of its 40 short sections lie with their mean turn on an ellipsoid and that the number of ampere turns per unit axial length is constant.

Performances. Tne main parameters and performances of eight solenoidal lens spectrometers are collected in Table I. The transmissions are in all cases very good, but the source size is distressingly low.

Additional references on focusing in uniform fields are given in (17-22).

Table 13.I

Performances of some solenoidal spectrometers

Authors	Length, cm	Source diameter, cm	Resolution, %	Transmission, %	Luminosits x 10³ cm²
Witcher	90	2	6	1	31
Haggstrom	90	0,5	2,5	2	4
Feldman, Wu	34	2	9	0,31	10
Schmidt	60	0,07	0,39	2	0,3
Bretonneau, Moreau	100	0,8	1,5	1,3	6,5
Jungerman et al.	228	0,22	0,024	0,24	0,09
Du Mond	63	0,6	1,3	4	10
Dolmatova, Kelman	45	/0,1 x 0,1 cm²/	1,4	5,2	0,52

References Ch.13

1. E. Persico, Rev. Sci. Instr. <u>20</u>, 191 (1949)

2. J. W. M. DuMond, Rev. Sci. Instr. <u>20</u>, 160, 616 (1949)

3. J. W. M. DuMond, Annals of Physics <u>2</u>, 283 (1957)

4. P. Hubert, C. R. Acad. Si <u>230</u>, 1464 (1950), Physica <u>18</u>, 1129 (1952), Ann. Phys. <u>8</u>, 662 (1953).

5. J. A. Jungerman, M. E. Gardner, C. G. Patten and N. F. Peek, Nucl. Instr. Meth. <u>15</u>,'1 (1962)

6. K. A. Dolmatova and V. M. Kelman, Nucl. Instr. Meth. <u>5</u>, 269 (1959).

7. J. P. Blewett, J. Appl. Phys. <u>18</u>, 968 (1947).

8. F. H. Schmidt, Rev. Sci. Instr. <u>23</u>, 361 (1952).

9. M. E. Gardner, J. A. Jungerman, P. G. Lichtenstein and C. G. Patten, Rev. Sci. Instr. <u>31</u>, 929 (1960).

10. H. Bosch, Publs. Comis. Nac. en. atom. Arg. No.2, 1956.

11. L. S. Goodman, Rev. Sci. Instr. <u>31</u>, 1351 (1960).

12. L. Snow, J. Res. Nat. Bur. St. C <u>69</u>, No 1, 49 (1965).

13. E. Hagström, Phys. Rev. <u>62</u>, 144 (1942).

14. L. Feldman and C. S. Wu, Phys. Rev. <u>76</u>, 180 (1949).

15. P. Bretonnean et J. Moreau, J. Phys. Rad. <u>14</u>, 25 (1953).

16. R. D. Birkhoff, A. W. Smith, H. H. Hubbell Ir. and J. S. Cheka, Rev. Sci. Instr. <u>26</u>, 959 (1955).

17. L. Dick, J. Phys. Rad. <u>17</u>, 590 (1956).

18. D. Beard, Rev. Sci. Instr. <u>28</u>, 19 (1957).

19. J. A. Jungerman and D. B. Beard, Rev. Sci. Instr. <u>27</u>, 56 (1956).

20. C. S. Wu, Phys. Rev. <u>115</u>, 108 (1959).

21. Koh, J. Phys. Soc. Jap. <u>4</u>, 245 (1949).

22. J. Mahauty, S. Pandit and C. Balakrishan, J. Sci. Ind. Res. <u>12B</u>, 571 (1953).

Text to Figures. Ch.13

Fig.13.1. A family of helical trajectories in a homogenous field.

Fig.13.2. The formation of the ring focus.

Fig.13.3. Graph of $\bar{\upsilon}_f$ as function of emission angle γ.

Fig.13.4. The definition of the ring focus by three principal trajectories.

Fig.13.5. Hubert baffle system.

Fig.13.6. Kelman's aberration corrector.

Fig.13.7. Du Mond's spectrometer with ellipsoidal coil for production of uniform field.

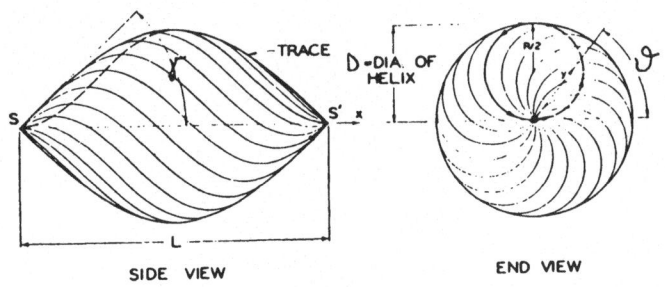

13.1

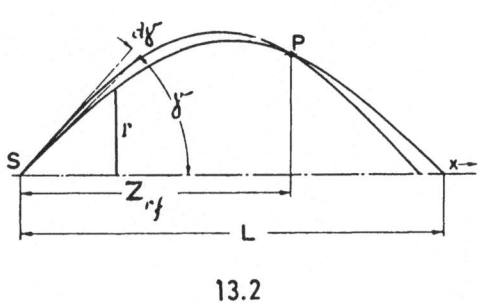

13.2

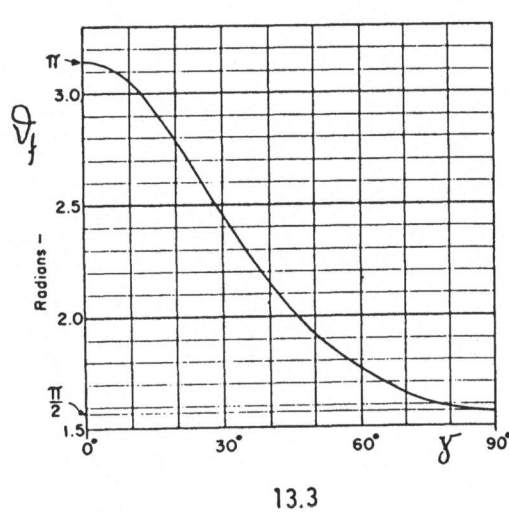

13.3

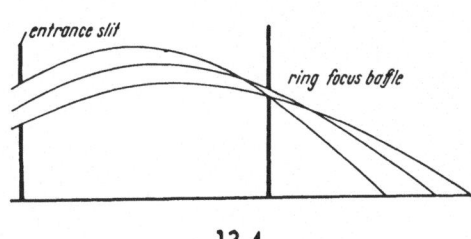

13.4

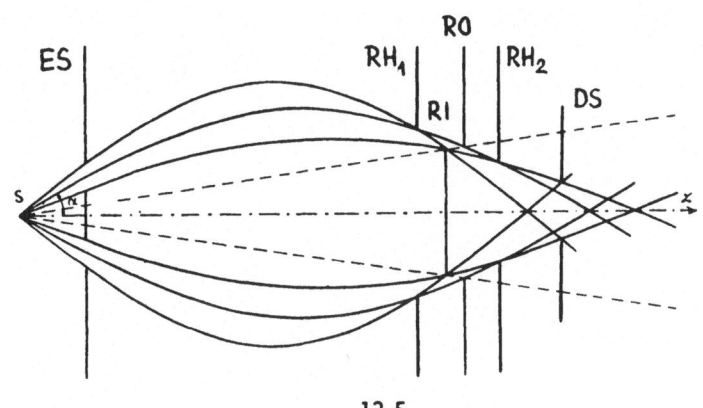

13.5

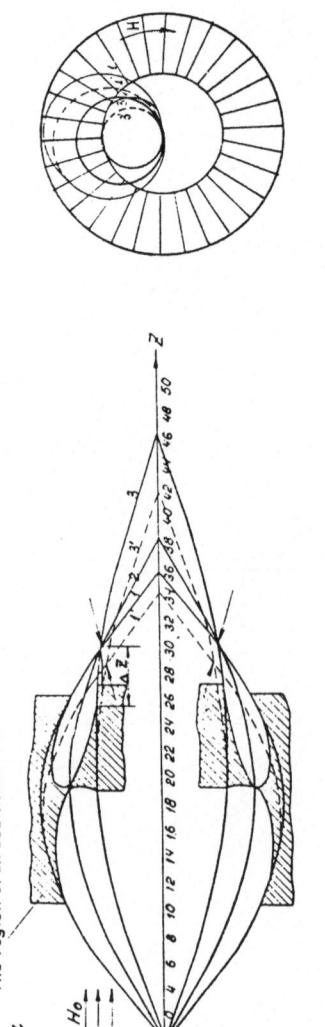

The region of an additional field

13.6

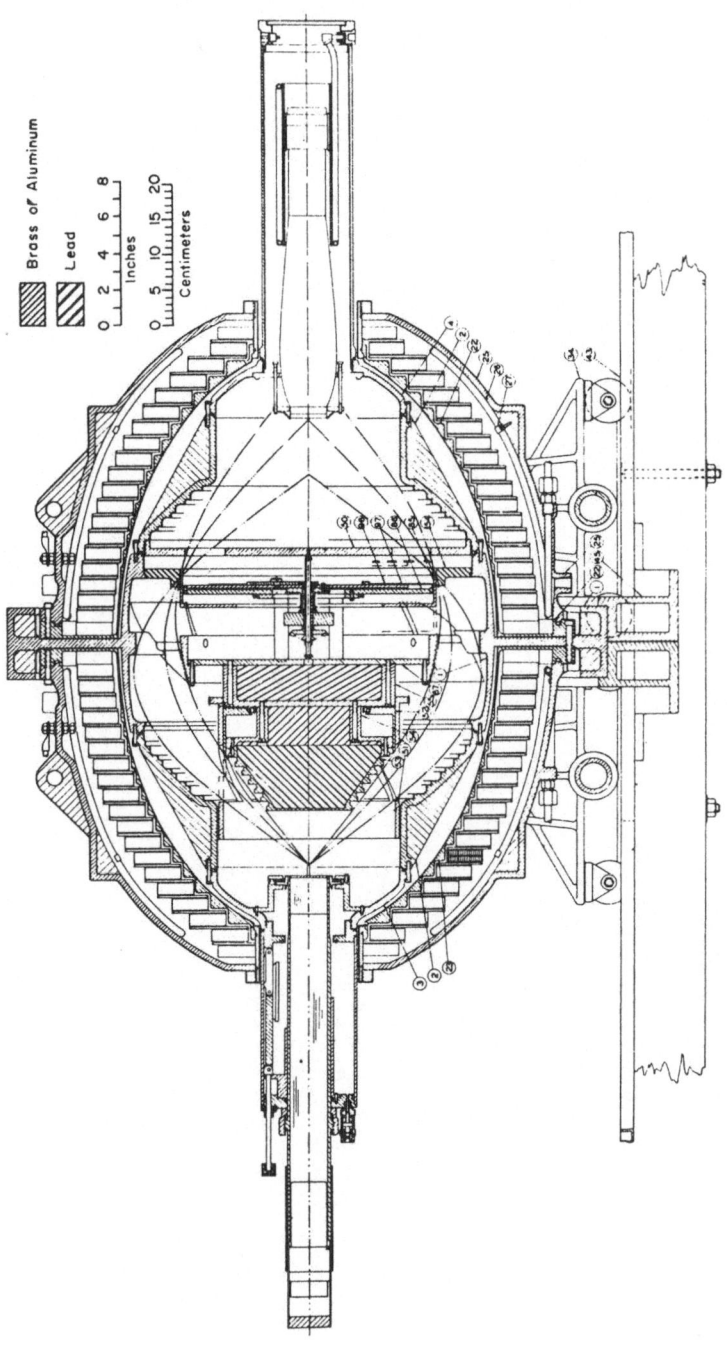

Brass of Aluminum

Lead

Inches

Centimeters

13.7

14. SHORT LENSES

The short lenses were popular, because of their simplicity, in the first decade following the World war II. By 1960's they have been mostly replaced by spectrometers with better performances, and built only for very specific purposes.

The short lens was introduced in beta-spectroscopy by Klemperer (1). First detailed investigations were made by Deutsch et al (2, 3) and by K. Siegbahn (4). Several spectrometers were subsequently described (5-11, 20). Ring focus was introduced at about the same time, as in solenoidal spectrometers (12-18), including the Hubert baffle (19). Dzhelepov and collaborators (21) have shown that toroidal correcting coils can substantially eliminate the spherical aberrations.

In the following we shall derive the focusing properties of a thin lens in paraxial approximation and give the performances of some short lens spectrometers.

<u>Use of axial field.</u> The calculation of electron lenses is facilitated by the possibility of presenting the components $B_r(r, z)$ and $B_z(r, z)$ of cylindrically symmetric fields, as functions of the field $B_z(o, z)$ on the symmetry axis. The relation between the field off and on the axis can be obtained by starting from the Laplace equation for magnetic scalar potential ψ, which in the case of cylindrical symmetry has the form

$$\frac{\partial^2 \psi}{\partial r^2} + \frac{1}{r} \frac{\partial \psi}{\partial r} + \frac{\partial^2 \psi}{\partial z^2} = 0 \qquad (14.1)$$

The solution of this equation can be represented by a series of powers of r, where the coefficients depend only on z

$$\psi(r, z) = \sum_{n=0}^{\infty} a_n(z) r^n \qquad (14.2)$$

Introducing (2) into (1) one obtains the relation

$$\sum_{n=0}^{\infty} \left[\frac{d^2 a_n}{dz^2} r^n + n^2 a_n(z) r^{n-2} \right] = 0 \qquad (14.3)$$

By equating to zero the coefficients of equal powers of r, a general reccurence relation follows from (3)

$$a_n(z) = - \frac{1}{n^2} \frac{d^2 a_{n-2}(z)}{dz^2} \tag{14.4}$$

and it is seen that the coefficients of odd powers of r are equal to zero. Once the coefficient $a_o(z)$ is known, all the others can be derived using the relation (4). The first coefficient a_o can be obtained by noting that at the symmetry axis, the series (2) reduces to the first term only

$$\psi(o, z) = \psi(z) = a_o \tag{14.5}$$

Using the reccurence relation (4), other coefficients can be derived, and (2) becomes

$$\psi(r, z) = \psi(z) - \frac{r^2}{4} \frac{d^2 \psi(z)}{dz^2} + \frac{r^4}{64} \frac{d^4 \psi(z)}{dz^4}$$

$$= \psi(z) - \frac{r^2}{4} \psi''(z) + \frac{r^4}{64} \psi^{IV}(z) \tag{14.6}$$

Since the components of the magnetic field are given by

$$B_z(r, z) = - \frac{\partial \psi(r, z)}{\partial z} \tag{14.7}$$

$$B_r(r, z) = - \frac{\partial \psi(r, z)}{\partial r} \tag{14.8}$$

and the relation (8) is reducing for the field on the axis to

$$B(z) = - \frac{\partial \psi(z)}{\partial z} \quad,$$

The general field components become

$$B_z(r, z) = B(z) - \frac{r^2}{4} B''(z) + \frac{r^4}{64} B^{IV}(z) - .. \tag{14.9}$$

$$B_r(r, z) = - \frac{r}{2} B'(z) + \cdot \frac{r^3}{16} B''(z) - \cdot \cdot \tag{14.10}$$

The vector potential A(r, z) can also be expressed in terms of the field on the axis, starting from Stokes theorem applied to vector potential

$$(\vec{A}\ \vec{ds}\ ) = \iint\limits_{S}\ (\text{curl }\vec{A})_n\ dS \tag{14.11}$$

where S is the surface of a circle, with the center on the symmetry axis and contained in a plane perpendicular to the axis. For cylindrically symmetric field (11) becomes

$$\int\limits_{0}^{2\pi} A_\theta\ rd\theta = \int\limits_{0}^{r}\int\limits_{0}^{2\pi}\ (\text{curl }\vec{A})_z\cdot rdrd\theta$$

$$= 2\pi \int\limits_{0}^{r}\ B_z(r,z)\ rdr \tag{14.12}$$

Integrating left side, one obtains

$$A_\theta \equiv A = \frac{1}{r} \int\limits_{0}^{r}\ B_z(r,z)\ rdr \tag{14.13}$$

Putting in the series expansion of B_z given by (9), the vector potential becomes

$$A(r,z) = \frac{r}{2}\ B(z) - \frac{r^2}{16}\ B''(z) + \ldots \tag{14.14}$$

Paraxial ray equation. Paraxial ray equation represents a simplified form of equations of motion valid for short lens and small emission angles. The simplifications follow from the following assumptions:

1. The particles come from a field free region where

$$B = 0 \qquad A = 0 \tag{14.15}$$

2. The aperture angle γ. which particle makes with the field axis is very small, so that

$$\cos\gamma \simeq 1 \qquad \sin\gamma \simeq \gamma \tag{14.16}$$

3. As electron enters the lens the angle between the velocity and the z-axis becomes even smaller, so that it can be taken that over the whole trajectory

$$v_z = \frac{dz}{dt} = v \cos \gamma \approx v = const \qquad (14.17)$$

4. It follows from (17) that in the lens the particle remains at approximately the same radial distance from the axis. It can be assumed then that

$$r \approx r_a = const \qquad (14.18)$$

5. Since γ and r_a are very small, higher orders of r can be neglegted in the series developments (9) and (14), giving

$$B_z(r,z) \approx B(z) \qquad A(r,z) \approx \frac{r}{2} B(z) \qquad (14.19)$$

6. In a short lens it can be taken that principal planes coincide. Fig.1 shows the parameters of a typical orbit. Particle is emitted at an angle γ_1, enters the lens at the plane A, remains in the lens at an approximately constant distance r_o from the axis, leaves the lens at the plane B and is focused at an angle γ_2.

Starting from the equations of motion for the cylindrically symmetric field

$$\frac{d^2 z}{dt^2} = -\frac{e}{m} r \frac{d\theta}{dt} \cdot \frac{\partial A}{\partial z} \qquad (14.20)$$

$$\frac{d^2 r}{dt^2} - r\left(\frac{d\theta}{dt}\right)^2 = -\frac{e}{m} \frac{d\theta}{dt} \cdot \frac{\partial (rA)}{\partial r} \qquad (14.21)$$

$$\frac{d}{dt}\left(r^2 \frac{d\theta}{dt}\right) = \frac{e}{m} \frac{dr}{dt}(rA) \qquad (14.22)$$

Equation (8) can be integrated immediately to

$$r^2 \frac{d\theta}{dt} - \frac{e}{m} rA = r_o^2 \left(\frac{d\theta}{dt}\right)_o - r_o A_o = C \qquad (14.23)$$

Since the particle starts from a point on the axis in the field free region and has no azimuthal component, the initial conditions are

$$r_o = 0 \quad A_o = 0 \quad \left(\frac{d\theta}{dt}\right)_o = 0 \quad C = 0 \qquad (14.24)$$

and (23) reduces to

$$\left(\frac{d\theta}{dt}\right) = \frac{e}{m}\frac{A}{r} \tag{14.25}$$

The vector potential $A(r,z)$ can be expressed by the field on the axis $B(z)$ using approximation (19), so that (25) becomes

$$\frac{d\theta}{dt} = \frac{e}{2m}B(z) \tag{14.26}$$

Using once more (19), the product rA becomes $(r^2/2)B(z)$ and

$$\frac{\partial(rA)}{\partial r} = rB(z) \tag{14.27}$$

With (26) and (27) the radial equation of motion (21) can be expressed as

$$\frac{d^2r}{dt^2} + \frac{e^2}{4m^2}B^2(z)\,r = 0 \tag{14.28}$$

The approximation (17) can be used to change the variables from t to z and (28) then becomes

$$\frac{d^2r}{dz^2} + \left(\frac{e}{2mv}\right)^2 B^2(z)r = 0 \tag{14.29}$$

This is paraxial ray equation. Using approximation (18), the integration between A and B gives

$$\left(\frac{dr}{dz}\right)_B - \left(\frac{dr}{dz}\right)_A = -r_a \left(\frac{e}{2mv}\right)^2 \int_A^B B^2(z)\,dz \tag{14.30}$$

Since $B(z)$ is different from zero only between A and B, the integration limits can be conveniently extended to infinity. Noting that

$$\left(\frac{dr}{dz}\right)_B \approx \frac{r_a}{b} \quad \text{and} \quad \left(\frac{dr}{dz}\right)_A \approx \frac{r_a}{a} \tag{14.31}$$

(30) becomes

$$\frac{r_a}{b} - \frac{r_a}{a} = -r_a \left(\frac{e}{2mv}\right)^2 \int_{-\infty}^{+\infty} B^2(z)\,dz \tag{14.32}$$

or

$$\left(\frac{1}{b}\right) - \left(\frac{1}{a}\right) = -\left(\frac{e}{2mv}\right)^2 \int_{-\infty}^{+\infty} B^2(z)\, dz \qquad (14.33)$$

By definition the focal length of the lens f is obtained when particle comes from the infinity. Then, for $a \to \infty$, $b = f$ is given by

$$\frac{1}{f} = -\left(\frac{e}{2mv}\right)^2 \int_{-\infty}^{+\infty} B^2(z)\, dz \qquad (14.34)$$

or expressing the momentum in $(B_o\rho)$ units

$$f = -4(B_o\rho)^2 / \int_{-\infty}^{+\infty} B^2(z)\, dz \qquad (14.35)$$

The rotation of the image is obtained from the equation (16), which with help of (17) can be written as

$$\frac{d\theta}{dz} = \frac{e}{zmv} B(z) \qquad (14.36)$$

The integration gives

$$\theta = \theta_A - \theta_B = \frac{1}{2(B_o\rho)} \int_{-\infty}^{+\infty} B(z)\, dz \qquad (14.37)$$

It follows from general properties of the equation (28) that all the rays starting from a point are again focused into a point. The image of a point on the axis is also on the axis, while a point off the axis is imaged into another point off the axis. This is demonstrated in many books on electron optics.

By comparing (33) and (34) the lens formula of geometrical optics is obtained

$$\frac{1}{a} + \frac{1}{b} = \frac{1}{f} \qquad (14.38)$$

Other formulae, discussed in Chapter 3. can also be deduced. We should only mention that the magnification M is given by

$$M = \frac{b}{a}$$

Spherical aberrations.   The paraxial approximations (15-19) cease
to be valid for non-negligible emission angles γ . An electron emitted at a
larger angle γ is focused at a shorter distance, as shown in Fig.2. The stronger focusing of the outer rays is due to two effects:

1. Outer rays have a smaller axial component of momentum. In the
limiting case of $\gamma = 90^{o}$, there could not be any axial motion at all.

2. In the center of a short lens, where the axial field is maximum,
the field off the axis increases with the radius, as can be seen from the first
two terms of (9),

$$B_z(r, z) = B(z) - \frac{r^2}{4} \, B''(z)$$

A typical short lens field is convex in the center, producing a negative $B''(z)$.
Electrons passing at larger radii experience therefore a stronger field and are
more strongly focused. This is very clearly shown in the theoretical investigations of Lindgren, which shall be discussed in the next chapter.

The spherical aberrations can be reduced either by having a concave field in the center or by additional correctors. The concave field will be
discussed in the next chapter.

Two kinds of correcting coils were used. Persico (22) proposed a
dipole magnet along the axis. Easier to calculate are the toroidal coils which
were developed in two Leningrads Laboratories, by Kelman and Dolmatova for
solenoidal spectrometer, as discussed in previous chapter, and by Dzhelepov
and collaborators for short lenses.

Aberration corrector. Dzhelepov, Tchan and Tiskin (21) designed
a toroidal correcting coil (Fig.3) made of 16 "slices". The section of the coil
is triangular and the shape was found semi-empirically. The coil is placed
near the source, where the field is very weak. The current in the corrector is
in series with the main coil.

Fig.4. shows the improvement obtained with the corrector. With a
source of 0,2 cm diameter, and the solid angle of 0,5%, the half-width without
corrector is 3,2%. With the corrector, the line becomes 2,5 times narrower

and twice higher. The improvement is quite eoncouraging.

Table I shows the performances of some short lens spectrometers.

Table 14.I

Performances of some short lens spectrometers

Authors	Length	Source diameter, cm	Resolution R, %	Transmission	Luminosity
Deutsch, Elliot, Evans	100	0.5	6	0.8	1.6×10^{-3}
		0.4	1.7	0.13	1.5×10^{-4}
Pratt, Boley	102	0.5	1.75	0.6	8×10^{-4}
Verster	100	0.3	1.3	1	7×10^{-4}
Dzhelepov, Tchan, Tishkin	100	0.2	3.2	0.5	1.5×10^{-4}
			1, 3 (corrector)		

References

1. O. Klemperer, Phil. Mag. $\underline{20}$, 545 (1935)

2. M. Deutsch, Phys. Rev. $\underline{59}$, 684 (1941)

3. M. Deutsch, L. G. Elliot and R. D. Evans, Rev. Sci. Instr. $\underline{15}$, 178 (1944)

4. K. Siegbahn, Ark. Mat. Astr. Fys. $\underline{28A}$, No. 17 (1942)

5. N. F. Verster, Appl. Sci. Res. BI $\underline{363}$ (1949)

6. E. N. Jensen, L. J. Laslett and W. W. Pratt, Phys. Rev. $\underline{75}$, 548 (1949)

7. G. Trumpy, JENER Report $\underline{10}$ (1952)

8. M. K. Banerjee and A. K. Saha, Proc. Phys. Soc. $\underline{1366}$, 937 (1953)

9. Z. Pleiner, Cheskoslov. Chasop. Fys. $\underline{5}$, 204 (1955)

10. T. Nainan, H. G. Deware and A. Mukerjec, Proc. Ind. Acad. Sci. A $\underline{44}$, 111 (1956)

11. M. Legros and G. Gueben, Bull. Scient. Assoc. Ing. $\underline{72}$ No. 2, 183 (1959)

12. W. W. Pratt, F. I. Boley and R. T. Nichols, Rev. Sci. Instr. $\underline{22}$, 92 (1951)

13. J. M. Keller, E. Koenigsberg and A. Paskin, Rev. Sci. Instr. $\underline{21}$, 712 (1950)

14. J. F. Perkins and A. W. Solbrig, Jr., Rev. Sci. Instr. $\underline{22}$, 173 (1951)

15. D. K. Butt, Proc. Phys. Soc. $\underline{62}$, 151 (1949)

16. K. C. Mann and F. A. Payne, Rev. Sci. Instr. $\underline{30}$, 408 (1959)

17. N. F. Verster, Physica $\underline{17}$, 637 (1951)

18. S. Frankel, Phys. Rev. $\underline{73}$, 804 (1948)

19. P. N. Mukherjee, M. K. Pal, M. K. Banerjee and A. K. Saha, Indian J. Phys. $\underline{31}$, 531 (1957)

20. R. Nathe, J. Schintlmeister, H. Seidenfaden and R. Weibrecht, Exptl. Tehn. Phys. $\underline{9}$, 1 (1961)

21. B. S. Dzhelepov, No Sen Tchan and A. Tishkin, Izvestia Akad. Nauk SSSR, Ser. Fiz. $\underline{20}$, 947 (1956)

22. E. Persico, Rend. Lincei $\underline{7}$, 191 (1955)

Text to Figures. Ch.14

Fig.14.1. The parameters in an idealised short lens.

Fig.14.2. The stronger focusing of outer rays in a short lens.

Fig.14.3. The correcting coil for short lenses, designed by Dzhelepov, Tchan and Tishkin.

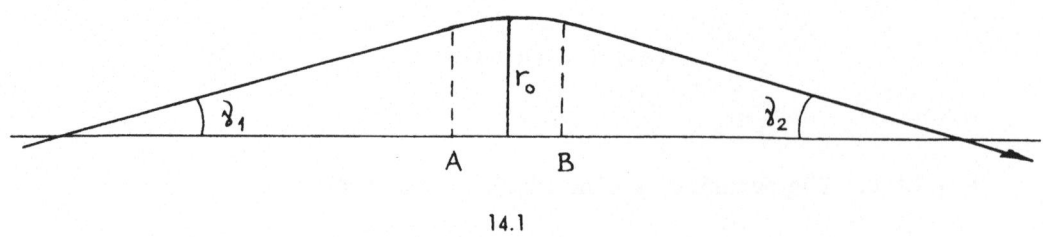

14.1

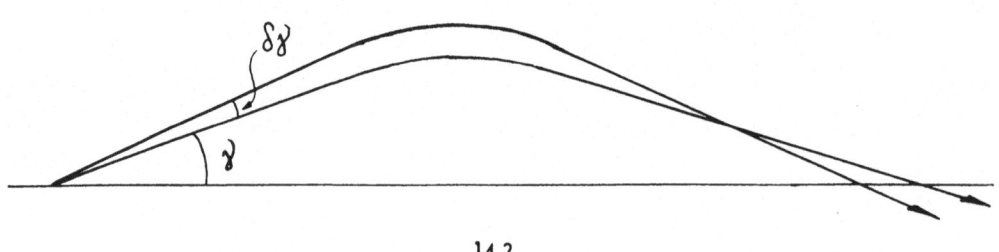

14.2

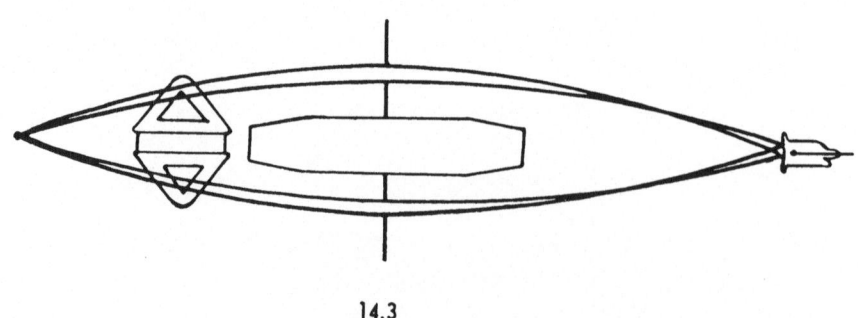

14.3

15. INTERMEDIATE LENSES

The construction of intermediate lenses followed the short ones (1-3), including the correcting coil, coaxial with the main coil (4-5). K.Sieg-bahn introduced the U-shaped field to reduce the spherical aberrations (6) and subsequently Slätis and Siegbahn developed the intermediate image spectro-meter (7-8), achieving high transmission with a small image. The intermedia-te image spectrometer was built in quite a few laboratories (10-14) and become commercially available (9).

Several spectrometers producing the U-field with separated lenses were made (15-26).

In the following we shall present the results of a theoretical study of intermediate types of lenses and their comparison to the solenoidal and short lenses.

Focusing properties. The focusing properties of 37 different field shapes were studied by Lindgren et al (28-30). The axial field shapes of the investigated fields are shown in Fig.1. As can be seen, most of the lens field shapes used so far are represented in the collection.

Two main contributions to the line width, from opening angle and the source size, were calculated separately. In all cases Hubert ring-focus baffle system was assumed, consisting of one inner and a series of outer baffles. The spherical aberration is measured by the ring-focus width Δr for the point source, which is given by

$$\Delta r = (\partial^2 r / \partial \gamma^2) \ \varepsilon^2 / 2 \qquad (15.1)$$

$$= (\partial^2 r / \partial \gamma^2) \ \Omega^2 / \sin^2 \gamma$$

and the base width, for a point source is then

$$R_p = \Delta r / p \ (\frac{\partial r}{\partial p})_{z_f} \qquad (15.2)$$

where $(\partial r / \partial p)_{z_f}$ is the momentum dispersion at the ring focus. Point source resolution is roughly proportional to the square of the solid angle and a conve-nient magnitude for the purpose of comparison is R_p / Ω^2.

The contribution of the source size can be investigated independently by considering trajectories of electrons emitted from the points along the periphery of the disc source, all at the same angle γ , corresponding to zero transmission. The width of such a beam I varies along the lens in a manner illustrated on Fig.2. The width of the beam has a minimum I_m at a distance z_I which generally does not coincide with point source ring-focus position z_f. The width of the beam at the point source ring focus I_f is then generally larger than I_m. If the ring-focus were placed at z_f, with the slit width $\Delta r + I_f$, the base width R_b would be

$$R_b = R_p + R_s \qquad\qquad (15.3)$$

where

$$R_s = I_f/p \ (\partial r/ \partial p)_{z_{f, \gamma}} \qquad\qquad (15.4)$$

The optimum position of ring-focus does not have to be necessarily z_f, and depends on relative contribution of Δr and I. For a large source and small γ, ring focus should be nearer to z_I, while for small source and large solid angle z_f would be better place. More precise information can be obtained by considering the trajectories emitted from all points of the source at $\gamma \pm \epsilon$. The radius of the source is s.

The relation (3) represents an upper limit of the base width and should be considered only as a rough indication. Since R_s is roughly propotional to s, a convenient magnitude for comparison is R_s/s.

The main results of the investigations made by Lindgren et al. are presented in Figs 3-4 and can be summarised in the following way:

- Point source resolution depends much on the field form and R_p/ Ω^2 varies by two orders of magnitudes between optimum fields and those having very large spherical aberrations.

- The minimum of the relative beam width I_m/s does not depend appreciably on the field form. The fields with smallest spherical aberrations have also smallest I_m/s.

– The minimum beam width I_m is almost always smaller than the source diameter $2s$. The ratio I_m/s decreases with emission angle γ and for $\gamma < 20^o$ can be smaller than 0.1. For the case of homogenous field $I_m = I_f = 2s$.

– The relative beam width at the ring focus I_f/s depends critically on the field form and can be up to one order of magnitude larger than I_m/s. The ratio I_f/I_s is largest for fields with smallest spherical aberrations.

– The characteristic behaviour of I_f/s is reflected in variation of R_s/s with the field form, which is also largest for fields with smallest spherical aberrations.

– The opposite behaviour of R_p/Ω^2 and R_s/s does not permit a straightforward selection of the "best" field shape.

– The focusing properties of some field shapes appear to be inferior even without further, more complex investigations. Such is the case of short lenses, which have large spherical aberrations producing large R_p/Ω^2, while I_m/s is also larger than in case of long lenses and I_f is approaching I_m.

– Maximum luminosity is obtained for homogenous field and the field No 24.5, which decreases in the middle to 75% of the value at the source and the detector.

– When specific activity is so high that small source size can be tolerated a   large solid angles are needed, the fields concave upwards with minimum ir.   e middle depressed to $25-50\%$ become more adequate.

References Ch.15

1. R. L. Graham and O. Klemperer, Proc. Phys.Soc. B 65, 921 (1952)

2. S. Jnanananda, J. Sci. Ind. Res. India 2, 397 (1952)

3. P. Hubert, J. Phys. Rad. 12, 763 (1951)

4. P. Hubert, C. R. Acad.Sci. 233, 943 (1951)

5. V. S. Shpinel, Zhur. Eksp. Teor. Fiz. 22, 255 (1952)

6. K. Siegbahn, Phil. Mag. 37, 162 (1946)

7. H. Slätis and K. Siegbahn, Ark. Fysik 1, 339 (1949)

8. H. Slätis, Sci. Tools 3, 12 (1956)

9. L. Wengstedt, Sci. Tools 4, No 2, 20 (1957)

10. G. Bolla, S. Terrani and L. Zappa, Nuovo Cim. 12, 874 (1954)

11. Z. Plajner and V. Brabec, Chekosl. Chas. Fys. A 10, 115 (1960)

12. K. Teuvo, P. Heikki and V. Heikki, Suom. tied. toim. Sar AVI, No 89
    (1962)

13. P. Depomier, M. Chabre, J. Crancon and H. Vialet, J.Phys.Rad 21, 493
    (1960)

14. T. Nagarajan, M. Ravindranath and K. Venkata Reddy, Nucl.Instr.Meth.
    67, 77 (1969).

15. E. A. Quade and D. Halliday, Rev.Sci.Instr. 19, 234 (1948)

16. H. M. Agnew and H. L. Anderson, Rev.Sci.Instr. 20, 869 (1949)

17. C. M. Van Atta, S. D. Warshaw, J. J. L. Chen and S. J. Taimuty, Rev.Sci.
    Instr. 21, 986 (1950)

18. C. S. Snowdon, Rev.Sci.Instr. 22, 878 (1951)

19. P. Grivet, J. Phys. Rad. 12, 1 (1951)

20. T. Azuma and K. Tsumori, J. Phys. Soc. Japan 7, 341 (1952)

21. G. P. Rundle, J. Ellis, T. C. Griffith and H. S. Tomlinson, Proc. Phys.
    Soc. B 67, 52 (1954)

22. H. Daniel and W. Bothe, Z. Naturforsch. 9a, 402 (1954)

23. K. Ohira, H. Matsui, K. Takazawa, M. Higuchi and F. Fukuzawa, Sci.
    Repts. Hyogo Univ. Agric.Ser. Nat.Sci. 2, 15 and 24 (1956)

24. K. P. Mitrofoaov and B. S. Shpinel, Izvestia Akad. Nauk SSSR, Ser. Fiz.
    21, 1607 (1957)

25. H. Daniel, Z. Naturforsh. 12a, 940 (1957)

26. W.Schneider, Z. Instrumentkunde 65, 103 and 126 (1957)

27. Y.Ramberg and A.B.Blauground Rev.Sci.Instr. 28, 286 (1957)

28. I.Lindgren and W.Schneider, Nucl.Instr.Meth. 22, 48 (1963)

29. I.Lindgren, G.Petterson and W.Schneider, Nucl.Instr. Meth. 22, 61 (1963)

30. I.Lindgren, B.Olsen, G.Petterson and W.Schneider, Nucl. Instr.Meth. 41, 331 (1966)

Text to Figures. Ch.15

Fig.15.1. The axial field shapes investigated by Lindgren et al.

Fig.15.2. The variation of source image I along the lens.

Fig.15.3. Disk source resolution at the ring focus for a point source versus
the point source resolution for some typical magnetic fields and
for different emission angles. Since the total resolution is roughly
composed of point and disk source contributions, the best field
shapes are generally those having both of them small.

Fig.15.4. The resolution vs the effective transmission for field 24.5 (a, b,
c, d), field 25 (e, f, g, h), field 11 (i, j, k, l) obtained by the
MC-method. Different positions of the outer baffles (a,c); (e,g);
(i, k). Different positions of the inner baffles (b, d); (f, h); (j,i).
Source radius 1 and 2 mm (a-h). Source radius 3 and 5 mm (i-l).

1000
800
600
400
200
B
z
25
50
2
23
27
32
20
8
28
33
UF
24
29
34
21
25
30
24.5
22
26
31
11

15.1

Image width
cm
.05
.04
.03
.02
.01
FIELD 2
α = 20°
s = 05 cm
0
10
20
30
40
50 cm
z
Source position
Ring-focus position
Detector position

15.2

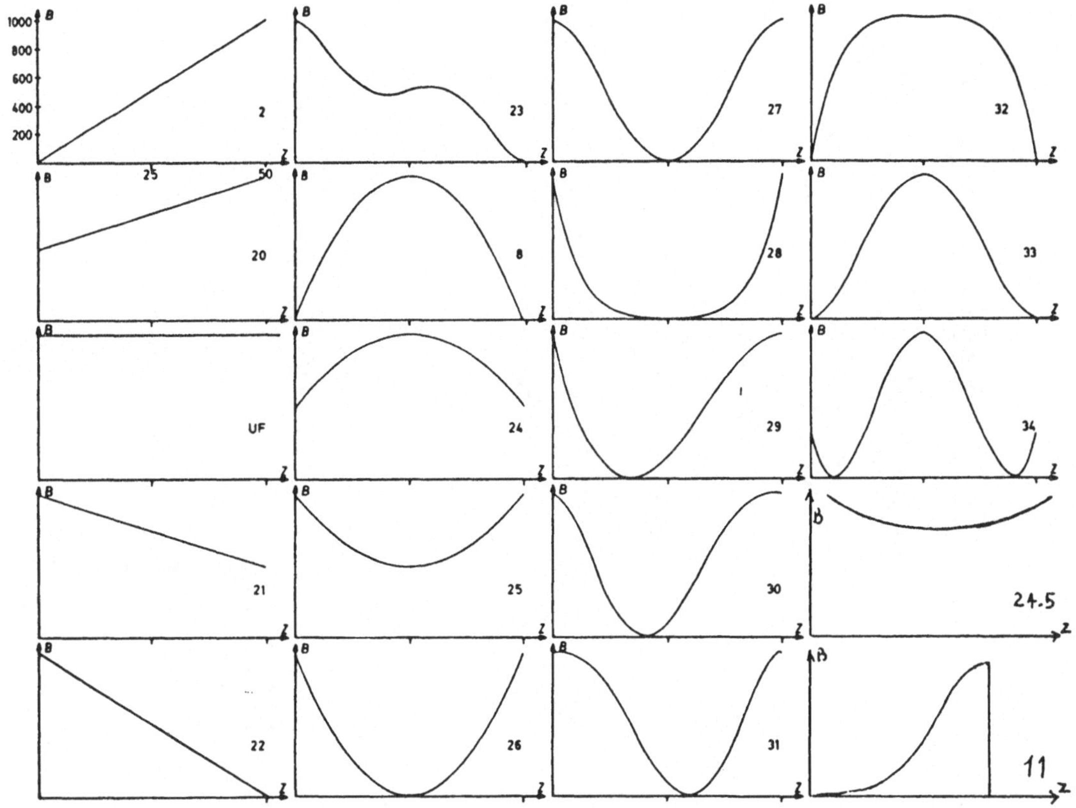

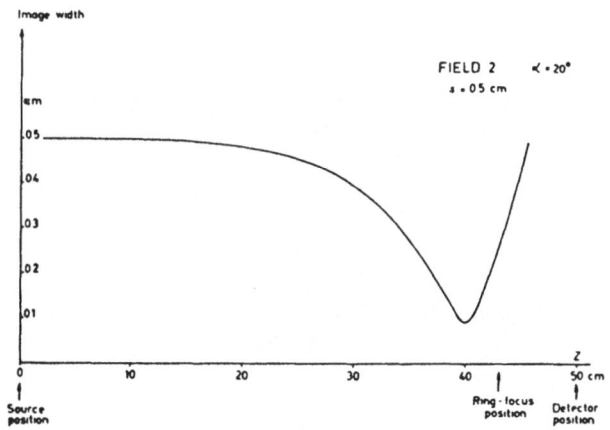

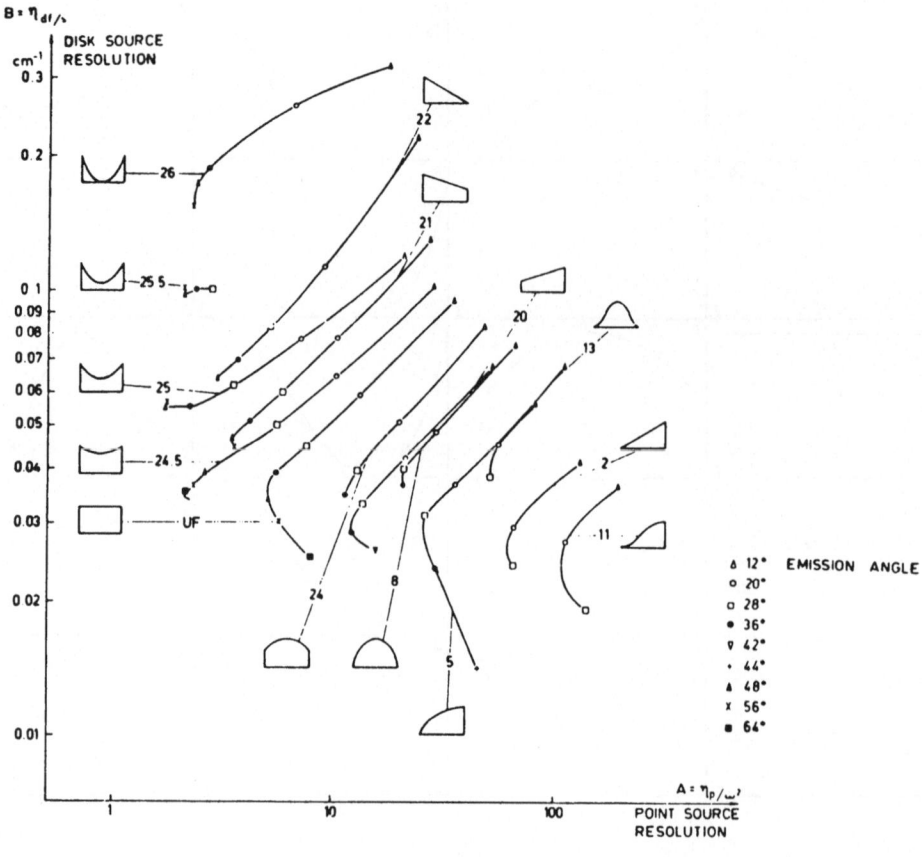

15.3

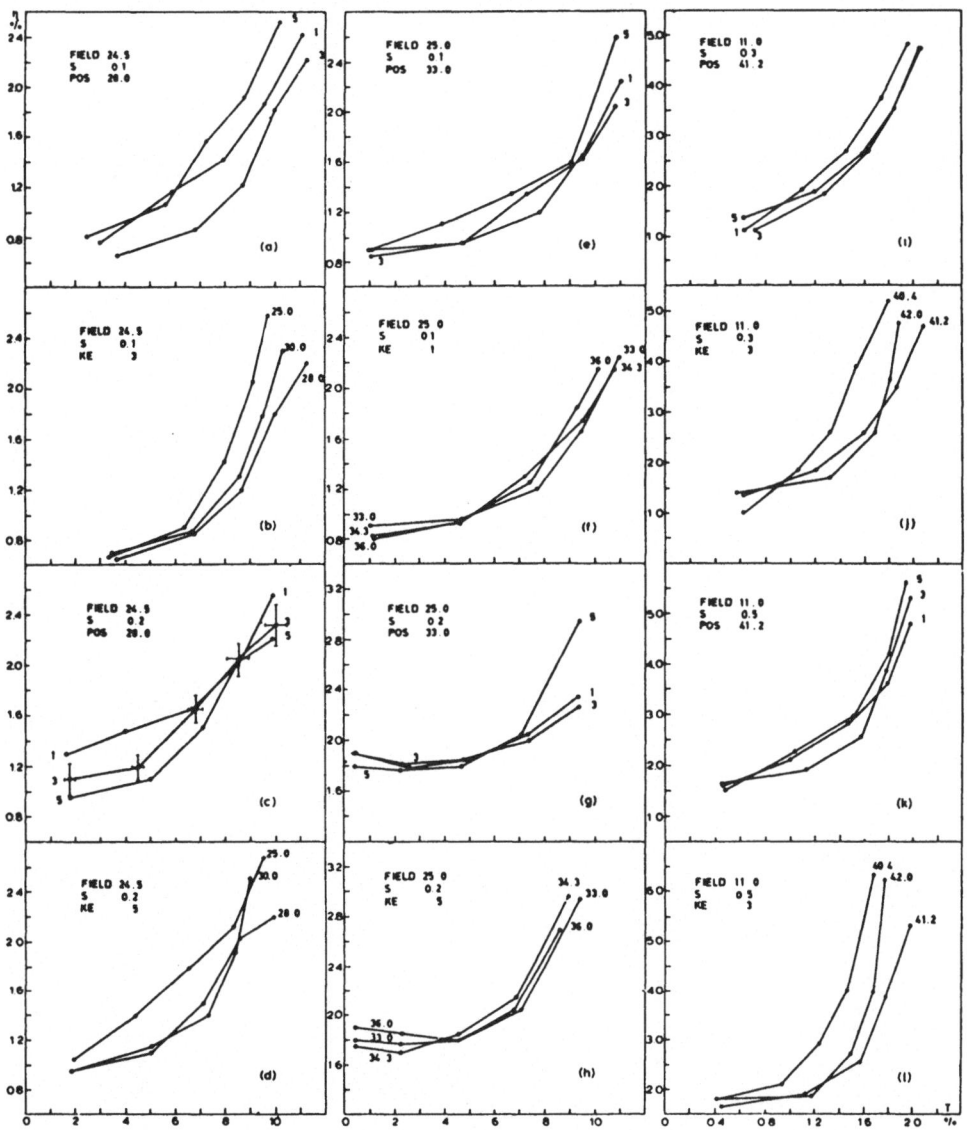

15.4

16. OPTICAL ANALOGY SPECTROMETER

Kelman and collaborators (2-8) developed an electron spectrometer analogous to the optical spectroscope, consisting of two lenses and a dispersive prism in-between. (Fig.1). The first lens collimates electrons into a parallel beam, which is deflected by the prism and focused by second lens. This is so far unique beta-ray spectrometer in which focusing and dispersion are separated and performed by different electron optical elements. The argument of authors (1) is that "this permits the use of elements most suited for each of these functions. Particles are focused by means of electron magnetic lenses which perform double focusing and whose aberrations are of the third order of magnitude with respect to the beam width. The deflecting field does not cause an additional line broadening".

The spectrometer based on optical analogy was first proposed and constructed by Klemperer (2). Although the spectrometer was sufficiently large (source-detector distance nearly 2 m long), the half-width of the F line was 15%. Kelman and Kaminskii have shown that the design of prism was not correct, being such that the focusing lens destroys the dispersion produced by the prism. They pointed out that the beam should remain parallel after leaving the prism and that a rectangular deflecting magnet, with a homogenous field, could be used for that purpose.

In such a magnet the parallel beam is focused into a line focus along x-axis (see Fig.1.a) and is emerging as a parallel beam again (if aberrations are neglected).

The source is rectangular. Since the image rotates in lenses, the source is not placed in vertical position, but is turned through an angle equal in magnitude to the angle of image rotation in the collimating lens, but in the opposite direction. For the same reason the exit slit is turned in the direction of the image rotation, so that it becomes parallel to the image.

Action of the magnetic prism

The deflecting magnet which acts as a prism has the following characteristics:

- The field is two dimensional, not depending on the x-coordinate.

- The action of the magnet may be described by dividing its field in three parts: central part with homogeneous field, and two fringing fields.

- Central part with homogenous field is responsible for dispersion.

- Fringing field acts as a cylindrical thin lens. A beam paralel to pole faces, entering at an angle α_1, is focused by first cylindrical lens into a line focus lying in the median plane mid-way between the sides of the magnet. The action of the second lens is symmetrical, so that a parallel beam leaves the magnet at the same angle α_1.

We shall first find the expression for the focal length of the cylindrical lens and then calculate the dispersion of the prism.

Cylindrical thin lens.- Before entering the lens electron is moving in x-y plane, its momentum making an angle α with y-axis. When $\alpha \neq 0$ the velocity has a component v_x which feels the B_y component of the fringing field, producing a force in z - direction. This force directs electron either towards the median plane, or away from it.

The z-coordinate of the electron is described by equation

$$m_o \ddot{z} = \frac{e}{c} \dot{x} B_y \qquad (B_x = 0) \qquad (16.1)$$

The component B_y can be developed in series near median plane, neglecting all powers of z higher than one. Since $B_y = 0$ for $z = 0$, B_y may be expressed as

$$B_y = z \left(\frac{\partial B_z}{\partial z} \right)_{z=0}$$

$$= z \left(\frac{\partial B_z}{\partial y} \right)_o \qquad \text{from rot } B = o$$

$$= z \frac{dB(y)}{dy} \qquad (16.2)$$

where $B(y)$ is the field in median plane.

The equations (1) and (2) give

$$\ddot{z} = \frac{e}{m_o c} \dot{x}z \frac{d\,B(y)}{dy} \tag{16.3}$$

In order to solve the equation (3), two approximations can be made:

- Within the thin lens, z-coordinate does not change much, so that one can take $z \approx z_o$,

- the entering angle α_1 does not change appreciably, if the thickness of the lens is much smaller than the radius of curvature inside the prism. One can take then $\alpha \approx \alpha_1$ and $\dot{x}$ becomes

$$\dot{x} = v \sin\alpha$$

$$\approx v \sin \alpha_1$$

Equation (3) becomes with these approximations

$$\frac{d^2z}{dt^2} = \frac{ev \sin \alpha_1}{m_o c} z_o \frac{d\,B(y)}{dy} \tag{16.4}$$

Using the relation $v = ds/dt$ equation (4) transforms into

$$\frac{d^2z}{ds^2} = \frac{e \sin \alpha_1}{m_o cv} z_o \frac{d\,B(y)}{dy} \tag{16.5}$$

Integrating (5),

$$\frac{dz}{ds} = \frac{e \sin \alpha_1}{m_o cv} z_o \int_{s_o}^{s} \frac{d\,B(y)}{dy} \cdot ds \tag{16.6}$$

The following approximation can also be made

$$\frac{dy}{ds} = \cos\alpha \approx \cos \alpha_1 \tag{16.7}$$

Noting tha the field at s_o is zero and at s can be taken to have the value B_m equal to the homogenous field inside the prism, the integration of (6) gives

$$\frac{dz}{ds} = \frac{e \, tg \, \alpha_1}{m_o cv} z_o B_m$$

$$= tg \, \alpha_1 \, z_o / \rho \qquad \rho = \frac{m_o cv}{eB_m} \qquad (16.8)$$

Leaving the lens, electron enters the homogenous region at an angle determined by dz/ds. The focal length f, measured from the side end of pole-pieces, is given by

$$f = \frac{z_o}{dz/ds}$$

$$= \frac{\rho}{tg \, \alpha_1} \qquad (16.9)$$

For $\alpha_1 < 90^o$ the lens is converging, the focal length being shorter for higher angle and higher field (smaller ρ).

There is only one angle α_1, for which the focus is falling midway between the pole ends. It is given by relation

$$\sin \alpha_1 = f$$

$$= \frac{\rho}{tg \, \alpha_1}$$

or

$$\sin \alpha_1 = \cot \alpha_1$$

$$\alpha_1 = 52^o$$

Since the whole treatment is approximate, this is also a very rough value. Integrating the equation of motion in the actual prism field the value of $\alpha_1 = 58^o$ was obtained and used.

Dispersion

To simplify the derivation we neglect the fringing field and suppose that the field is confined to the space between the pole pieces falling abruptly to zero outside of it.

An electron moving in x-y plane and entering the deflector at an angle α_1 leaves the deflector at an angle α_2, as shown in Fig.2. Within the magnet, electron describes a segment of a circle with radius ρ. If the momentum of the electron were larger by a very small amound Δ (B ρ), the radius of curvature would become $\rho + \Delta\rho$, and it would leave the deflector at an angle $\alpha_2 - \Delta\alpha_2$. The following equations connect the angles and the radius of curvature

$$\rho \sin \alpha_1 + \rho \sin \alpha_2 = A \qquad (16.10)$$

$$(\rho + \Delta\rho) \sin \alpha_1 + (\rho + \Delta\rho) \sin (\alpha_2 - \Delta\alpha_2) = A \qquad (16.11)$$

Developing (11) taking into acount that $\Delta\alpha_2$ and $\Delta\rho$ are small quantities, and subtracting (10) from it, one obtains

$$\frac{\Delta\alpha_2}{\Delta\rho/\rho} = \frac{\sin \alpha_1 + \sin \alpha_2}{\cos \alpha_2}$$

When $\alpha_1 = \alpha_2 = \alpha$ angular dispersion becomes

$$\frac{\Delta\alpha}{\Delta(B\rho)/(B\rho)} = 2 \, tg\alpha \qquad (16.12)$$

The linear dispersion D is obtained by multiplying the angular dispersion with focal length of the focusing lens.

$$D = 2 \, tg \, \alpha \cdot f_f \qquad (16.13)$$

Aberrations and resolution

Let us first consider the contribution of source width s to image width. Considering the rays in the median plane of an ideal prism, one can take that a small source width s would cause a change of entrance angle α, by an amount $\Delta\alpha_1$ equal to

$$\Delta\alpha_1 = s/f_c \qquad (16.14)$$

In the same way, rays leaving the prism with angles α_2 difering by $\Delta\alpha_2$ would cause a contribution to image width Δ_s equal to

$$\Delta_s = \Delta\alpha_2 \cdot f_f \tag{16.15}$$

It is easy to show that in the symmetric case when entering angle α_1 is equal to exit angle α_2, one obtains $\Delta\alpha_1 = \Delta\alpha_2$. When these angles are not equal, following equations describe the relations between angles for electrons of same energy, entering at angles α_1 and ($\alpha_1 - \Delta\alpha_1$), leaving at angles α_2 and ($\alpha_2 + \Delta\alpha_2$) respectively (Fig. 3).

$$\rho \sin \alpha_1 + \rho \sin \alpha_2 = A \tag{16.16}$$

$$\rho \sin (\alpha_1 - \Delta\alpha_1) + \rho \sin (\alpha_2 + \Delta\alpha_2) = A \tag{16.17}$$

For small $\Delta\alpha_1$ and $\Delta\alpha_2$ these equations give

$$\frac{\Delta\alpha_1}{\Delta\alpha_2} = \frac{\cos \alpha_2}{\cos \alpha_1} \tag{16.18}$$

For $\alpha_1 = \alpha_2 = \alpha$ $\Delta\alpha_1 = \Delta\alpha_2 = \Delta\alpha$

and the partial image width Δ_s becomes

$$\Delta_s = \Delta\alpha \cdot f_f$$

$$= s \cdot f_f/f_c \tag{16.19}$$

The ratio f_f/f_c represents the magnification.

Another contribution of the lenses to the image width Δ_{sp} is due to spherical aberrations. Parallel beams are focused into images of finite size, measured by diameters of confusion disks d_c and d_f. As it was the case with source width, d_c is magnified by the factor f_f/f_c, so that Δ_{sp} becomes

$$\Delta_{sp} = d_c \cdot f_f/f_c + d_f \tag{16.20}$$

The aberrations produced by the prism are mainly due to the action of cylindrical lenses. Finite width of the beam in vertical direction causes a broadening of linear focus in the middle of the prism, which in its turn produceds a spread of angles in the vertical direction of the beam leaving the prism. A point source forms a linear image parallel to the exit slit. This requires a higher exit slit, but does not contribute to the line width.

We will now find the aberrations produced by vertical extension of the source. Let us consider electron leaving the source at a distance z_1 from median plane, in direction making an angle α_1 with x-y plane. It leaves the prism at an angle α_2 (Fig.4). The projection of the momentum in field-free region will be $v \cos \gamma_1 \sin \alpha_1$, before entering the prism, and $v \cos \gamma_2 \sin \alpha_2$, after leaving the prism.

As the field does not depend on x-coordinate, the projection of the momentum on x-axis in the field-free region is constant. Denoting the vector potential $A(y,z)$ as A_1 and A_2 in two field-free regions, where vector potential is constant, one can write

$$m_o v \cos \gamma_1 \sin \alpha_1 + \frac{e}{c} A_1 = C \qquad (16.21)$$

$$m_o v \cos \gamma_2 \sin \alpha_2 + \frac{e}{c} A_2 = C \qquad (16.22)$$

Subtracting (22) from (21)

$$\cos \gamma_1 \sin \alpha_1 - \cos \gamma_2 \sin \alpha_2 = \frac{-e}{m_o vc} (A_1 - A_2) \qquad (16.23)$$

For an electron moving in median plane, (23) becomes

$$\sin \alpha_{1m} - \sin \alpha_{2n} = - \frac{e}{m_o vc} (A_1 - A_2) \qquad (16.24)$$

Subtracting (24) from (23)

$$\cos \gamma_1 \sin \alpha_1 - \cos \gamma_2 \sin \alpha_2 = \sin \alpha_{1m} - \sin \alpha_{2m} \qquad (16.25)$$

In our case the trajectories in the median plane are symmetric, the angles α_{1m} and α_{2m} are equal and denoted by α_1 so that (25) transforms into

$$\cos \gamma_1 \sin \alpha_1 = \cos \gamma_2 \sin \alpha_2 \qquad (16.26)$$

An electron leaving the line source at a point $z \neq 0$ will have $\alpha_1 = \alpha$ and in the prism it would be curved slightly more than an electron moving in the x-y plane, because only the projection of the momentum to the x-y plane is effective in Lorenz force action. If will leave the prism at an angle $\alpha_2 = \alpha + \Delta\alpha$. The equation (26) can be expressed as

$$\cos \gamma_1 \sin \alpha_1 = \cos \gamma_2 \sin (\alpha + \Delta\alpha) \tag{16.27}$$

Since γ_1 and γ_2 are small angles, one can develop in series $\cos \gamma_1$ and $\cos \gamma_2$. Equation (27) then becomes

$$\Delta\alpha = - \operatorname{tg}\alpha \left(\frac{\gamma_1^2}{2} + \frac{\gamma_2^2}{2} \right) \tag{16.28}$$

Passage through the second cylindrical lens is nearly symmetric with the passage through the first one, so that $\gamma_2 \approx - \gamma_1$, and (28) becomes

$$\Delta\alpha = - (\operatorname{tg}\alpha) \gamma_1^2 \tag{16.29}$$

To find the vertical coordinate z_2 and lateral coordinate η of the image, we should note that smallness of γ's allows the following relations

$$z_1 = - \gamma_1 f_c \tag{16.30}$$

$$z_2 = \gamma_2 f_f \tag{16.31}$$

which give for z_2

$$z_2 = \gamma_1 f_f$$
$$= z_1 f_f / f_c \tag{16.32}$$

The lateral coordinate η becomes

$$\eta = \Delta\alpha \cdot f_f$$
$$= - (\operatorname{tg}\alpha) \cdot \gamma_1^2 \cdot f_f$$

$$= -(\text{tg } \alpha) \cdot \frac{z_1^2}{f_c^2} \, f_f$$

$$= -(\text{tg } \alpha) \cdot \frac{z_2^2}{f_f} \tag{16.34}$$

Above results show that the image of a line source will be curved. If the slit is shaped accordingly, this should not contribute to the broadening of the image. The curvature is very small, since for $z_2 = 1$ cm, lateral displacement is of the order of $0,01$ cm.

When a spectrometer consists of three separate iron-clad magnets, one of them having a completely different geometry from the other two, the matching of fields requires great care, and any mismatching would contribute to the line broadening. If currents in the lenses differ by amounts ΔI_f and ΔI_c from values required to focus electrons deflected by the prism, the image is broadend by an amount Δ_ν given approximately by

$$\Delta_\nu = 2 d \left(\Delta I_f / I_f + f_f / f_c \cdot \Delta I_c / I_c \right) \tag{16.35}$$

where d is the width of the collimated beam. Since $2d \simeq 10$ cm, a mismatching of 1% would be equivalent to adding 1 mm to the width of the source, which is itself about 1 mm wide in the high resolution work. The matching is therefore very important.

The resolving power R may be written

$$R = \frac{\Delta_{im} + \Delta_{s\ell}}{2D} \tag{16.36}$$

where $\Delta_{s\ell}$ is the exit slit width and the image width Δ_{im} is equal to

$$\Delta_{im} = (s + d_c) \, f_f / f_c + d_f + \Delta_\nu \tag{16.37}$$

Deflecting magnet

The ideal field of deflecting magnet can be defined by following requirements:

- Field should be two-dimensional, not depending on x-coordinate.

- Left and right fringing flux distributions, which act as cylindrical lenses should be identical.

- Inside field should be homogeneous, without local variations.

- Symmetry with respect to median plane.

- Field should be everywhere perpendicular to x-axis.

- Field geometry defined by above requirements should be independent of field strength or of previous field history.

- There should be no stray fields.

These requirements are not simple and development of a satisfactory magnet has probably taken more time and effort than any other component of the spectrometer.

In principle, some of requirements can be relaxed. Thus, authors show that the two-dimensionality is not strictly needed provided that integral

$$\phi = \int_{-\infty}^{+\infty} B(\ell) \cos \alpha(\ell) d\ell \qquad (16.38)$$

remains constant, or in worse case changes linearly with x-coordinate. In the above integral $B(\ell)$ is the vertical field in the median plane, ℓ is the distance along the trajectory and $\alpha(\ell)$ is angle between an element $d\ell$ of trajectory and the y-axis. This requirement probably does not help much in the designing stage, but might be usefull in the final adjustment of a given magnet.

A schematic view of the deflecting magnet, cut by an y-z plane is shown in Fig. 5. To insure x-independence, the length (100 cm) is five times the beam width. The width of the poles is 50 cm and the height of pole gap 15 cm. The parallelism of pole pieces along x-axis was achieved within 0,003%.

The field variations along the x-axis do not exceed 0,01% over the region through which the beam is passing and for energy range of 11 to 2250 keV. The authors state (1) that the field geometry depends only slightly on previous state of magnetisation.

The lef-right symmetry was also achieved and is independent of the field strength in the working energy range.

One can see on Fig.5. the shields which cut the stray fields between the upper and lower parts of the yoke, that had to be made large in order to accomodate the beam.

Magnet is excited by the coils wound around the pole-pieces.

Magnetic lenses

Several different lenses were in use while the spectrometer was developed. We shall only mention latest type of constant geometry lenses and recently developed lenses with variable geometry.

Constant geometry lenses. A cross-section of the lens is shown in Fig.6. The coil is wound directly on the vacuum cylinder. The density of turns is highest in the middle, falling gradually towards the ends. The maximum focal length is 127 cm.

An iron cylinder placed over the winding serves as a shield, the field falling to zero at 20 cm from its edges.

The field distributions of two lenses was found to be identical within the measurement error of 0,3%.

An operating disadvantage of this type of lenses is that in order to change the solid angle, one has to move the source. This requires a readjustment of lens current and source rotation angle. In the symmetric field version, the focal length could be changed from 127 to 28 cm, changing thereby the solid angle from 0,04% to 0,8% of 4 π.

Variable geometry lenses. Peregud, who continued the development of this type of spectrometer, when Kelman left Leningrad, designed with collaborators (10) lenses with variable geometry. They consider that fixed geometry lenses had two important disadvantages:

- The increase of solid angle is limited by the requirement that the focal length of the collimating lens cannot be made shorter than the half--width (at half-height) of the bell-shaped field distribution. Making lens shorter does not pay, because spherical aberrations increase.

- If the focal length of collimator lens is decreased, while f_f remains the same, magnification f_f/f_c increases prohibitively, for large solid angles.

The variable geometry lens is shown in Fig. 7. with corresponding field distributions. The source is placed in the center of the lens, so that only one half is used. The unused half of the lens is cut at 10 cm from the center and the field is not symmetrical. By switching in variations parts of the coil, the focal length can be set at various values, changing the solid angle from $0,06$ to $3,6\%$ of 4π . The lens is 240 cm long.

Focusing lens is identical to the collimating one. This offers two possibilities. Magnification can be kept equal to one. The second advantage is that for high resolution, lenses with wide field distribution are used, which have smaller spherical aberrations. The contribution of aberrations to the line width becomes then negligible as can be seen from Table I., giving the partial width of source slit and lens aberrations.

Table I

D, m	Ω , %	source		$\Delta_{s\ell}$, mm	detector slit $\Delta_{s\ell}/D\%$	aberrat. d/D %	R, %
		s, mm	s/D %				
4,4	0,06	0,4	0,008	0,4	0,008	0,0001	0,016
3	0,17	1	0,033	1	0,033	0,003	0,07
1,3	0,7	1	0,077	1	0,077	0,027	0,18
0,75	2	1	0,13	1	0,13	0,183	0,45
0,55	3,6	1	0,18	1	0,18	0,33	0,7

Performances

The tipical performances of earlier spectrometer with fixed geometry are given in Table II.

Table II

f_c	source, mm^2	slit width mm	R	Ω	L_g, cm^2
127	0,4 x 15	0,4	0,014	0,0045	3.10^{-6}
127	0,4 x 15	0,4	0,022	0,012	7.10^{-6}
127	1 x 15	1	0,036	0,04	6.10^{-5}
73	2 x 15	2,5	0,1	0,1	3.10^{-4}
42	1,5 x 15	2,5	0,14	0,37	8.10^{-4}
28,5	1 x 15	2,5	0,2	0,79	$1,2.10^{-3}$

Spectrometer with variable geometry, which is now commercially available under marks UMB1 and UMB2 has following performances (UMB-1).

Table III

source, mm^2	slit, mm^2	R, %	Ω, %
		0,06	0,06
		0,09	0,17
1 x 15	1 x 15	0,14	0,34
		0,2	0,7
		0,25	1,1
		0,62	2
		0,77	3,6

These are obviously very preliminary results, which do not yet show the possibilites of the spectrometer.

Other spectrometers

Another spectrometer with fixed field was constructed in Bucharest (9), with performances similar to those in Table II.

A spectrometer was built with only one (collimating) lens (8), but the performances are inferior to two lens type.

Kelman and Yavor discuss in their book "Electron optics" (10)an optical analogy spectrometer with electrostatic deflector.

References Ch.16

1. O.Klemperer, Phil.Mag. 20, 545 (1935)

2. V.M.Kelman, B.P.Peregud and V.I.Skopina, Nucl.Instr.Meth. 27, 190 (1964)

3. V.M.Kelman and D.L.Kaminsky, Zh.Exptl.Teor.Fyz. 21, 555 (1951)

4. V.M.Kelman, D.L.Kaminsky and V.A.Romanov, Izv. Akad. Nauk SSSR, ser.fyz. 18, 148 (1954)

5. V.M.Kelman, D.L.Kaminsky and V.A.Romanov, Izv.Akad.Nauk SSSR, ser.fyz. 18, 209 (1954)

6. V.M.Kelman, B.P.Peregud and V.I.Skopina, Atomnaya Energiya, 10, No 5, 534 (1961)

7. V.M.Kelman, B.P.Peregud and V.I.Skopina, Zh.Exptl. Teor.Fyz.
   32, 1446 (1962)
   32, 1465 (1962)
   29, 1919 (1959)

8. V.M.Kelman, N.M.Dusaev, G.S.Malkiel and N.N.Nevodnitchny, Zh. Exptl. Teor. Fyz. 26, 107 (1954)

9. A.Gelberg, I.Ringhiopol, C.Protop, I.Vita, C.Tripa Revue Roumaine de Phys. 13, 375 (1968)

10. V.M.Kelman and S.Ya. Yavor, Electron optics, Akad.Nauk SSSR, 1963, Leningrad (in russian)

Text to Figures. Ch.16

Fig.16.1. Electron spectrometer based on optical analogy.

Fig.16.2. Paths in median plane traced by two electron having momenta $B\rho$ and $B\rho + \Delta(B\rho)$, both entering the prism at an angle equal to α_1.

Fig.16.3. Paths in median plane traced by two electrons having equal momenta, entering at angles α_1 and $(\alpha_1 - \Delta\alpha_1)$.

Fig.16.4. Path of an electron entering the prism at an angle γ_1 with respect to x-y plane.

Fig.16.5. Deflecting magnet.

Fig.16.6. Constant geometry lenses.

Fig.16.7. a) Variable geometry lens. b) Spectrometer with variable geometry lenses.

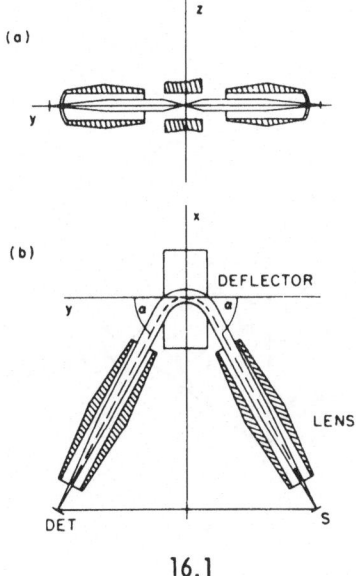

16.1

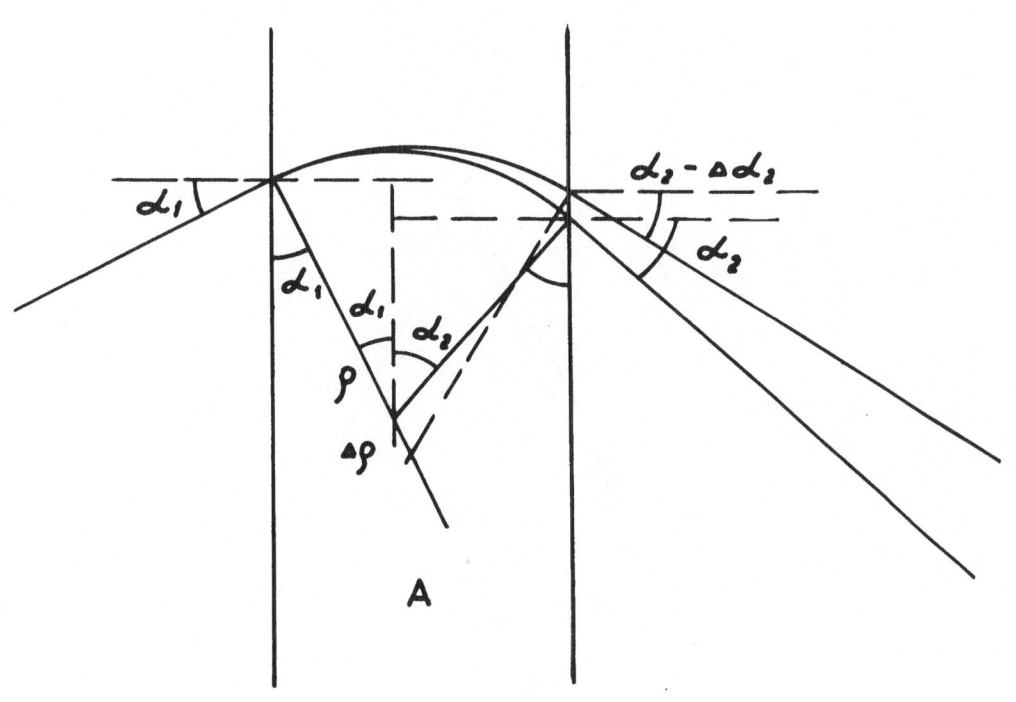

16.2

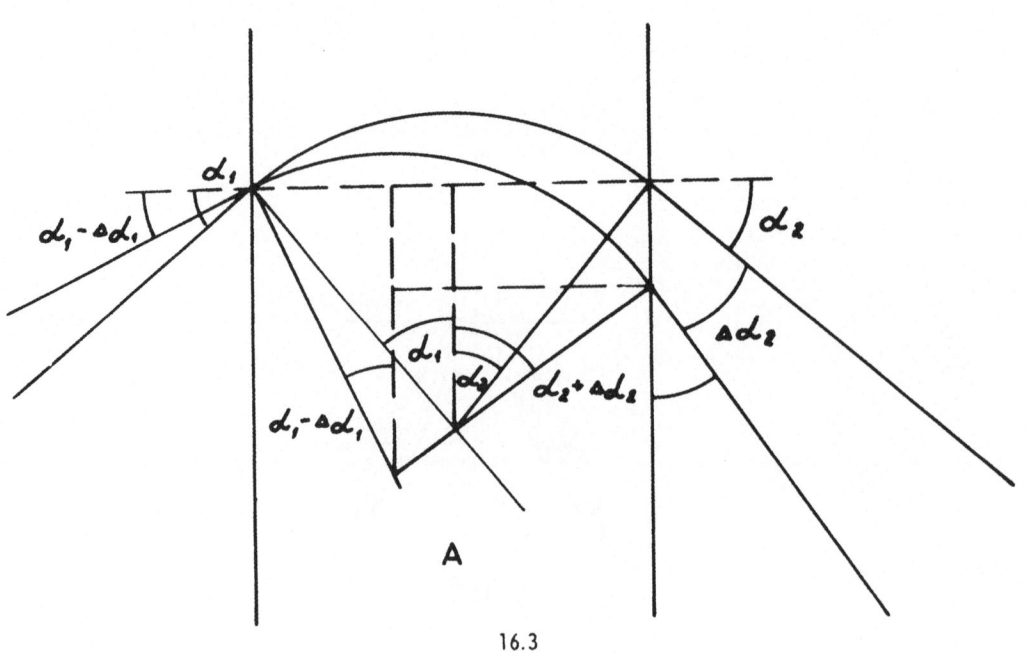

A

16.3

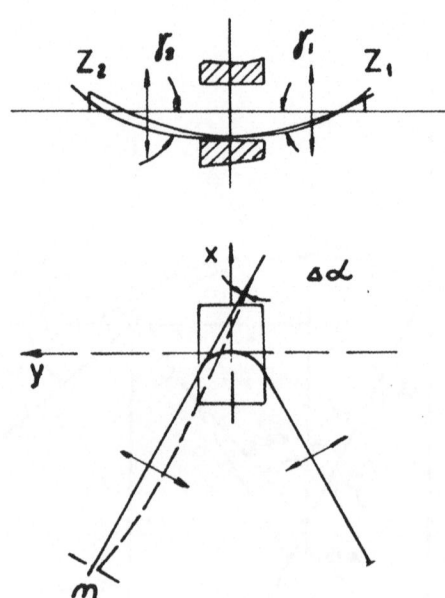

16.4

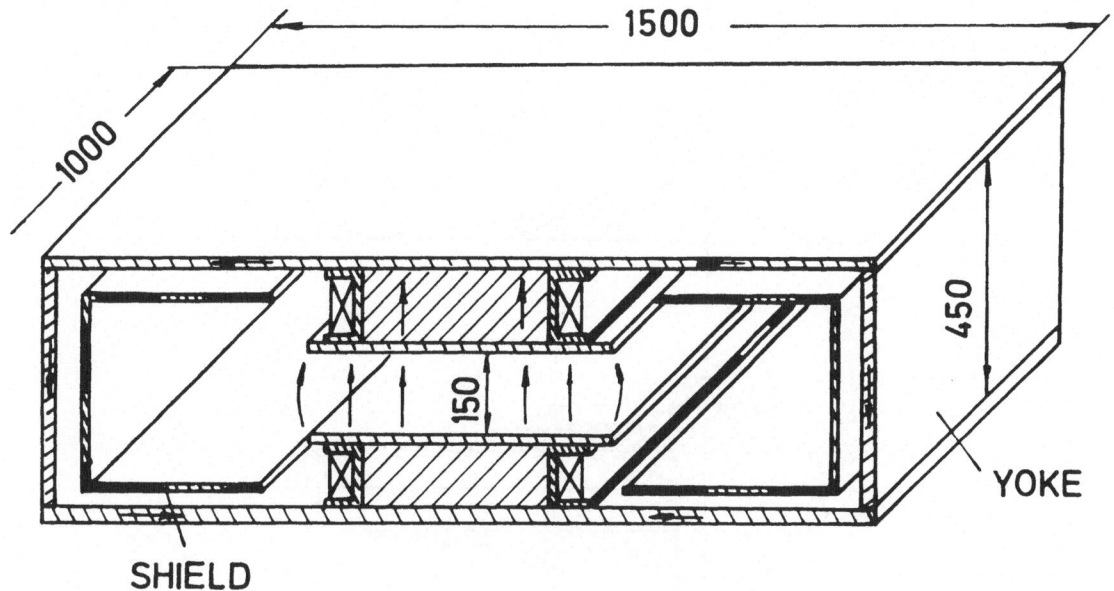

16.5

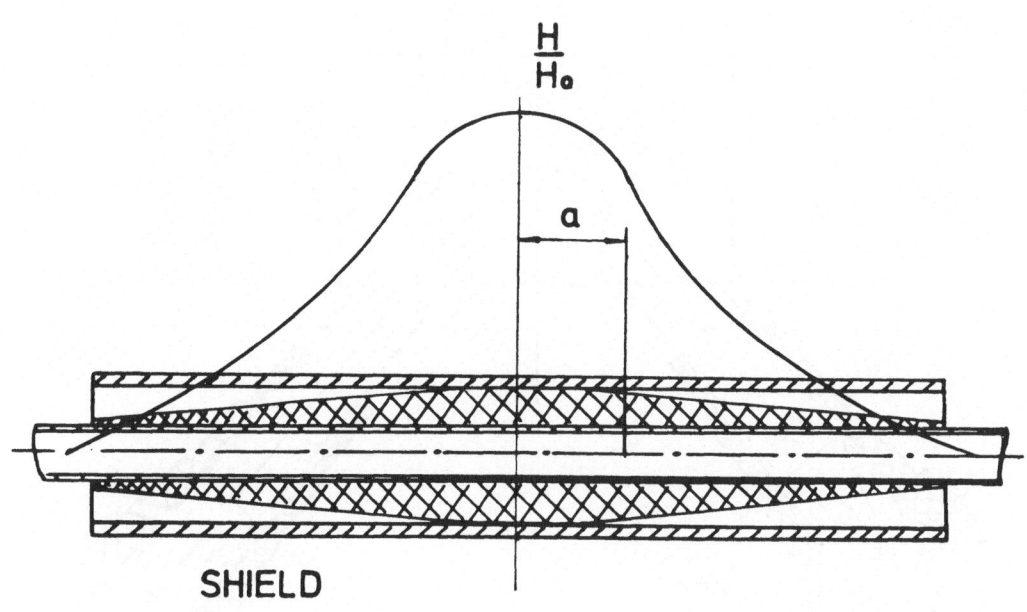

16.6

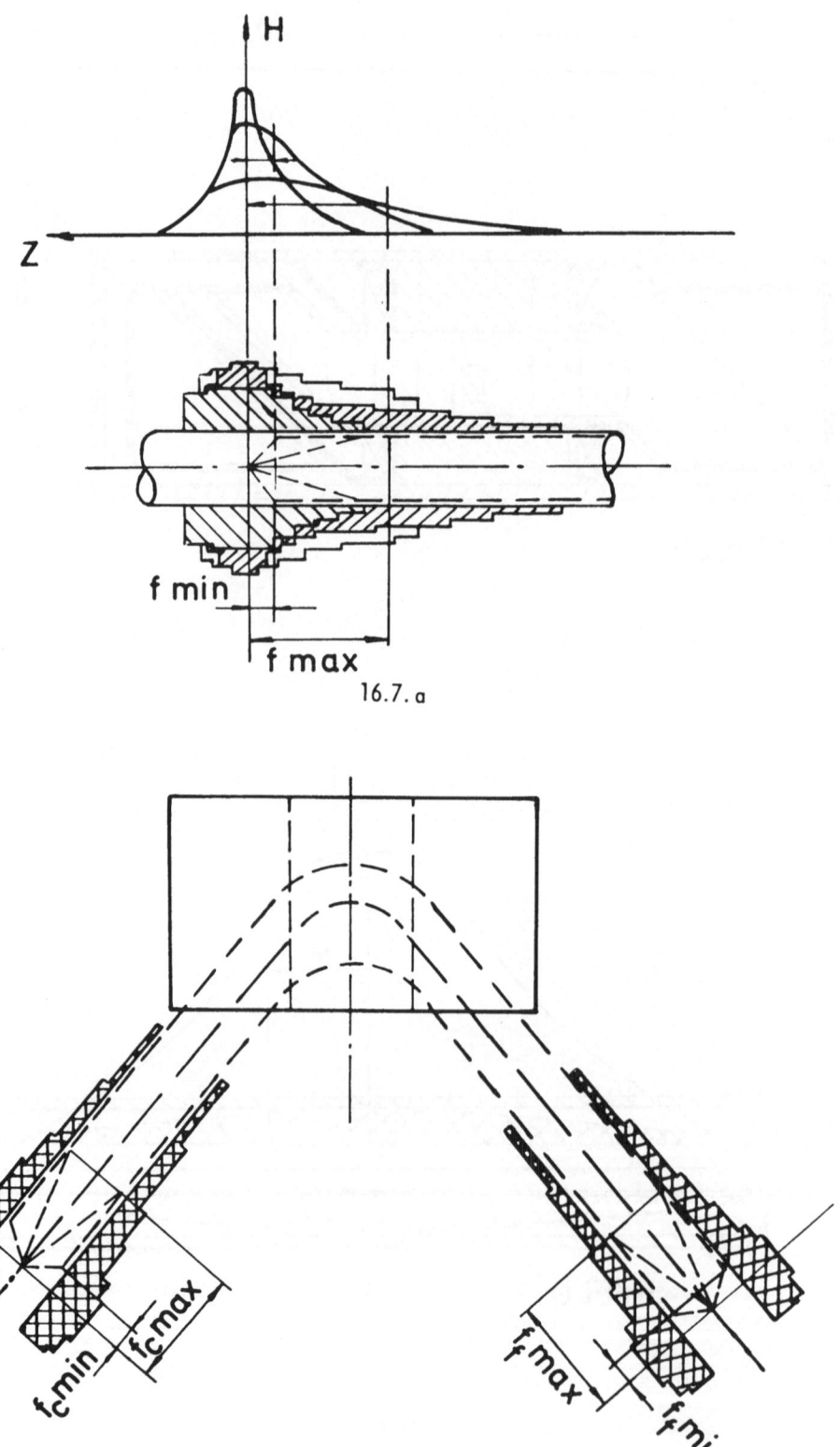

16.7. a

16.7. b

Selected Issues from

Lecture Notes in Mathematics

Springer-Verlag
Berlin
Heidelberg
New York

Lecture Notes in Physics